D1757033

# SHARPEY'S FIBRES

## THE LIFE OF WILLIAM SHARPEY
## THE FATHER OF MODERN PHYSIOLOGY IN
## ENGLAND

Front cover: *Sharpey's admission ticket to Joshua Brooke's Anatomy School, dated 11th June 1821.* (Courtesy of Angus Council Cultural Services)

Back cover: *Syllabus of the course of lectures at Joshua Brooke's School.* (Courtesy of Angus Council Cultural Services)

*1. William Sharpey in middle age. (Courtesy of the Boston Medical Library in the Francis A. Countway Library of Medicine, Boston, USA.)*

# Sharpey's Fibres

## The Life of William Sharpey the Father of Modern Physiology in England

*by*

Alan H. Sykes

William Sessions Limited
York, England

ISBN 1 85072 270 6

*Profits from the sale of this book are donated to the Sharpey Physiological Scholarship at University College London.*

Printed in 11 on 14 point Times Typeface
from Author's Disk
by Sessions of York
The Ebor Press York, England

# *Contents*

# List of Illustrations

# *Acknowledgements*

IT IS A PLEASURE to express my thanks to the librarians, archivists and others at the many organisations I have visited. I would mention particularly University College London, The Wellcome Library, The Royal Society, The Royal Society of Medicine, the New Dictionary of National Biography at Oxford University Press, the Signal Tower Museum at Arbroath, The Royal College of Surgeons of Edinburgh and the university libraries of Cambridge, Edinburgh, Glasgow and Lancaster. Such people seem to have been specially selected for their cheerful willingness to help aspiring authors. My apologies are due to the staff of those organisations which I have not been able to mention by name. My thanks for permission to reproduce photographs and extracts from other works are recorded where they appear in the text. I am particularly grateful to Bob Sissons of William Sessions Limited for his expert help in preparing the text.

Literary ambition, no matter how modest, extracts its price from domestic harmony and so to Margaret, who had to put up with the priority given to word processing, the frequent, long, visits to libraries, and the inevitable gloom when things were not going too well, to her, above all, I am most grateful.

# Introduction

IN ANY PHYSIOLOGICAL Hall of Fame one would expect to find a place for the names of, for example, William Harvey, Stephen Hales and Luigi Galvani. In more recent times the outstanding achievements of the French and German schools would be recognised by the inclusion of Francois Magendie, Claude Bernard, Hermann von Helmholtz and Carl Ludwig, among many others. But it is highly unlikely that the name of William Sharpey would be found even if the selection were restricted to British physiologists. This is perhaps surprising given that Sharpey has been widely acclaimed as "the father of modern physiology in England" a title perhaps more likely to be conferred upon Michael Foster or Burdon Sanderson, the founders of the Physiological Society. What, then, are the reasons for the comparative obscurity of Sharpey? It has to be said that he did not acquire his illustrious sobriquet by virtue of his physiological discoveries; in truth he never made any. He was an anatomist albeit a modern one who had taken up the microscope rather than the scalpel and who had promoted the new science of histology in the medical curriculum. It was with the microscope that he made his only original contribution to anatomy, the first description of the fibrous tissue in bone which later became known as Sharpey's fibres. Although he held a number of public positions, including that of Secretary of the Royal Society, he was never a statesman of science in the mould of Thomas Huxley or Lyon Playfair. He was a modest, unassuming man of little real scholastic achievement but one who had a passion for learning and for passing on his learning to others. He was a great teacher who inspired a number of talented young men to devote themselves to physiology as a career. Their names are synonymous with the renaissance of English physiology which, in its achievements, made it the equal of the great European schools. Their attachment to Sharpey as mentor and friend made them, in a sense, Sharpey's fibres; hence the title of this book.

The lives of the famous physiologists have been the subject of many books and papers but the great teachers tend to be overlooked (there is no English biography of Ludwig for example) even though their influence has often been profound. William Sharpey has a place in the history of physiology in England and it is the aim of this book to give an account of his life and of his part in the establishment of physiology as an independent discipline in Great Britain.

The stimulus to writing this book came from an invitation to contribute a revised entry on Sharpey for the new edition of the Dictionary of National Biography at present in preparation. The original entry in that British archive of the great and the good was written by D'Arcy Power in 1905 and it still provides an accurate and useful summary of Sharpey's life but since then more information has come to light. The scholarly account of his life and work by Taylor (1971) will long be an essential starting point for anyone with a serious interest in Sharpey or in physiology in nineteenth century England. The correspondence between Sharpey and his close friend Allen Thomson, as set out in the book by Jacyna (1989), adds more historical detail and a glimpse into the personalities of these two dedicated Scottish teachers. It also includes a discerning introduction and a wealth of biographical references. Without the help of these two important sources my task would have been immensely harder and I should like to acknowledge my debt to both authors. From the above and from other material which I had collected I was able to complete the new entry but owing to the necessary limitations of space much had to be omitted.

There is, however, more to Sharpey's life than his part in the emerging science of physiology; the academic world of his time was not only one of great altruism and dedicated scholarship but also of rivalry, jealousy and much ill-will. Although I would place him on the side of the saints rather than the sinners, (if such a simplistic division be possible) Sharpey was involved in some famous rows which were carried on with a ferocity unfamiliar today. His life therefore shows us something of the university and scientific world of mid-Victorian England.

Having come to know Sharpey I felt that others might wish to be introduced to his life and times and, since not everyone has access to academic journals or the inclination to consult them, I decided to put together a brief, illustrated life of Sharpey which would provide an easily accessible and readable account. To this end I have omitted footnotes but provided sufficient indication of my sources to enable the more dedicated reader to go into the subject in greater depth.

*2. The old Surgeons Hall, Edinburgh, built 1697 and used until 1832 when the new hall was built. It was purchased by the Infirmary and later passed to the University. From a drawing by Sandby, 1753.* (Courtesy of the Royal College of Surgeons of Edinburgh)

# SHARPEY'S FIBRES

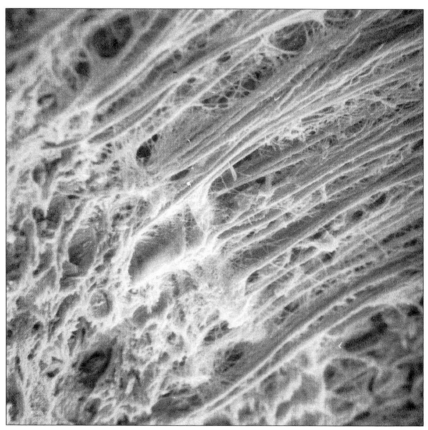

*3. Scanning electron micrograph of a muscle attachment site on a monkey humerus. Sharpey's fibres of uncalcified collagen bundles (upper right) have become mineralised as bundle bone (lower left).* (Illustration and notes kindly supplied by Sheila Jones, University College London)

# CHAPTER 1

## *Scotland*

ARBROATH IS A SMALL town situated on the east coast of Scotland between Dundee to the south and Aberdeen to the north in what was the county of Forfarshire, now included in the larger District of Angus, the older name for the region. It derives its name from being at the mouth, or aber in Gaelic, of the small river Brothock which enters the sea at this point. In the 18th and early 19th centuries this river provided water power for a number of textile mills spinning flax from jute imported through nearby Dundee and woven into sail cloth for the fleets of the world. At its peak there were over thirty spinning mills and also weaving shops and rope works. The town prospered and the port was developed for the coastal trade, then the fastest means of transport between parts of Britain, and for the expanding fishing fleet. Despite the coming of the railways in the 1830's, the port was enlarged and improved in 1844 and with it came boat building and repairing A striking architectural feature from those times is the Signal Tower built in 1813 as a signalling station to the Bell Rock lighthouse 12 miles out to sea. It is now a splendid museum illustrating the history of the town and its famous inhabitants. With industry came banks, churches, schools and all the urban amenities of an affluent and expanding community. New buildings in the local red sandstone were a source of civic pride and imparted a sense of warmth sometimes at odds with the biting winter winds which sweep in from the North Sea. Engineering and other trades grew and although sail gave way to steam, causing a decline in the textile industry, the town continued to be an industrial centre rather than one dependent upon fishing and agriculture. It was here, in 1841, that Alexander Shanks wrote his patent application for a lawn mower which he

1

developed into a world-wide business and the largest employer in the town. The population of only 2500 in 1742 grew to over 20,000 by 1890.

Arbroath owes its origins to the Abbey founded in 1178 by King William who granted the monks the right to establish a burgh with a port and a weekly market. It was one of the richest foundations in Scotland. The town received its Royal Charter from James VI in 1599 but the Abbey was dissolved at the Reformation, in 1608, and for three centuries was looked upon only as a source of high quality building stone. In later, more enlightened, times its ruins have been carefully preserved and today it attracts visitors from far and wide. The principal historical event associated with the Abbey was the letter written to the Pope by a number of Scottish noblemen in 1320 in which they affirmed their allegiance to the Scottish King, Robert the Bruce, and their independence from England. This, the Declaration of Arbroath, has been hailed as corner-stone of Scottish nationhood ever since.

It was at this both ancient and modern town that William Sharpey was born on the 1st of April 1802. His father Henry Sharpey was an Englishman from Folkestone, Kent, where the unusual name seems to have originated. He was a shipping agent with the firm of Cullen which sent him to open a branch in Arbroath in 1794. While working there he met and married Mary Balfour on Christmas Day 1795. She was born in 1774, the eighth child of David and Elizabeth Balfour, an old Arbroath family, and died in 1836. One of her brothers, David, was a merchant in the town and he often sent cargoes by sea, which might explain how Henry Sharpey came to know the family. David served in the militia and held the office of Provost. Mary was "a women of rare accomplishments" with a wide knowledge of the old ballad literature of Scotland; many of these works have been preserved as a result of her exceptional memory, a characteristic which she passed on to her son William.

She had five children: Elizabeth, her eldest, Isabella, Henry, David, and William her youngest. Henry and David died in their boyhood leaving William without the company of his older brothers but later he acquired step-brothers. His father had died, aged only 43, six months before William was born and since he himself had no children the surname in this branch of the family died out. His sister Elizabeth married William Colvill, a medical practitioner, and they had two children, Mary and William Henry, both of whom played a part in their uncle's later life. Mary, who never married, became his housekeeper in London until illness overtook her in 1872 and she returned to Arbroath where she died a few years later. William qualified as MRCS England in 1855 and became a medical officer in the Indian Army; his studies had been paid for by his uncle, an act of generosity which he never forgot. In a letter to Edward Schafer, many years later, he wrote: "Uncle Sharpey is all in all to me, all I live for and all I care for".

The other sister, Isabella, married a Major Goodall and died in Heidelberg; she did not maintain the same close links with her brother.

Sharpey's mother remarried in 1806 to Dr. William Arrott, a local medical practitioner from a respected family of long-standing in the area. He had studied at Edinburgh, where he obtained his diploma from the Royal College of Surgeons, and at St. Andrews where he took his MD. He died in 1862, aged 89, after a lifetime of service to the community.

They had six children, four sons and two daughters, the half-sibs of William who grew up in a lively family atmosphere in which education and culture played an important part. Of the sons of this marriage, three, Henry, James and David, became medically qualified and, after further study abroad, took up general practice in the area; the fourth, Alexander, did not enter medicine but he was involved with microscopy at one time and corresponded with William about it. There was, therefore, a family involvement with medicine, science and scholarship, a background which no doubt had its influence on William. The daughters, Mary and Jacobina, both married and kept up a correspondence with William, outliving him by several years.

Sharpey, as he will be referred to henceforth, kept in touch with his family all his life; he mentions them in many of his letters and often met them on his frequent visits to Scotland. He had cousins in Kent, the children of his father's brother William, but there is no record of any contact with them. A family tree of the Sharpey and Arrott families is given in the Notes. It has been derived from the International Genealogical Index and from dates found on gravestones and in various biographies but it can not be taken as authoritative since it has not been checked against primary sources such as parish registers.

Although he had an English father and spent most of his working life in England, Sharpey always regarded himself as a Scot; he had a deep love for his home town and his country and he never missed an opportunity to re-visit Scotland. He died in London but it was his wish to be buried in Arbroath and so, on the morning of the 15th April 1880, his coffin was taken from University College in a stately procession of over fifty private carriages of his friends and admirers, to Euston station for his final journey to Scotland. The funeral the next day was a private one but as a mark of respect most of the shops along the route were closed and the church bell tolled. The coffin was borne from the Abbey gate by pall bearers who included his dear friend Allen Thomson, the only mourner from London, his half-brother Dr. James Arrott of Dundee, the Provost of Arbroath and a number of medical men from the area who knew him. As reported in the British Medical Journal, "a large crowd of onlookers paid their respects to their distinguished townsman, whose past career had shed lustre on his birthplace, and whose personal qualities endeared him to all who knew him". His body was laid to rest in his family grave in Arbroath Abbey.

The gravestone is inscribed on both sides. The west side records his mother and her surviving children:

*Mary Balfour widow of Henry Sharpey*
*married Dr. Wm.Arrott of Arbroath*
*Died 1st May 1836 aged 62 years*
*and was interred near this spot*
*Elizabeth daughter of*
*Henry Sharpey and Mary Balfour*
*and wife of Wm. Colvill*
*died in London 26th April 1855 aged 58 years*
*and was interred in Highgate Cemetery*
*Isabella*
*also their daughter, wife of*
*Major George Goodall*
*died at Heidelberg 18th February 1861 aged 62 years*
*and was buried there*
*William Sharpey MD FRS*
*born 1st April 1802 died 11th April 1880*
*and was buried here*

*4. Arbroath harbour circa 1890 when it was a busy commercial port.*
*(Courtesy of Angus Council Cultural Services, PA1980.639)*

*5. Arbroath in the 19th century looking across the harbour towards the Signal*
*Tower. Note the barrels used for packing herrings for export. (Courtesy of Angus Council*
*Cultural Services, PA1979.953)*

*6. Arbroath harbour today.*

*7. Old houses in Arbroath which date from Sharpey's time.*

*8. Signal Tower Museum,
Arbroath, which houses
Sharpey artefacts.*

*9. Arbroath Abbey, front view.*

*10. Arbroath Abbey, graveyard. Abbey buildings would have formerly covered this site.*

*11. The Old College, Edinburgh University as it was in Sharpey's time.*
(Courtesy of the Wellcome Library, London)

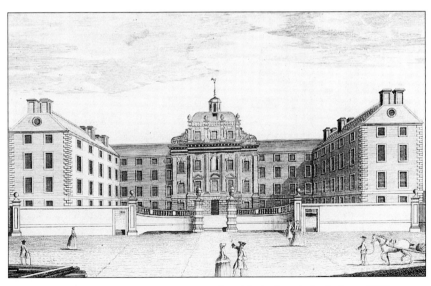

*12. The old Edinburgh Infirmary, opened in 1741, demolished about 1879.*
(Courtesy of the Royal College of Surgeons of Edinburgh)

13. *South-west corner of Surgeons Square, Edinburgh when Sharpey was a student. On the left is the old Surgeons Hall; on the right the hall of the Royal Medical Society; in the centre the building with the pillars contained the lecture rooms of the anatomists Barclay and Knox; in the background, between it and Surgeons Hall, is the house of John Thomson where Sharpey and Allen Thomson lectured. From a drawing by Shepherd, 1829. (Courtesy of the Royal College of Surgeons of Edinburgh)*

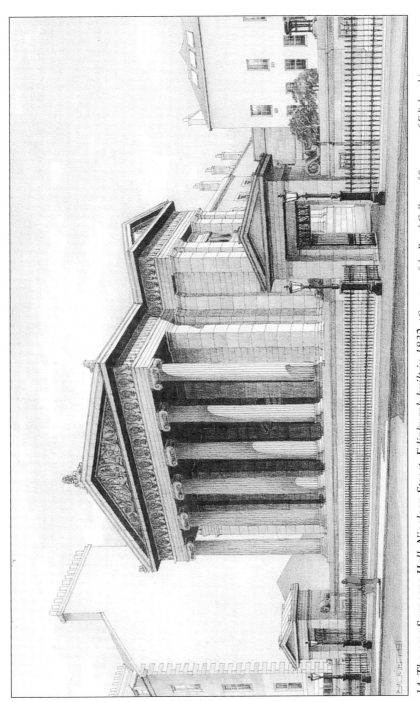

14. *The new Surgeons Hall, Nicolson Street, Edinburgh built in 1832. (Courtesy of the Royal College of Surgeons of Edinburgh)*

*15. Portrait of Thomas Morton (1813-1849) by his brother Andrew Morton.*
*(Courtesy of the Royal College of Surgeons of England)*

*16. Portrait of Allen Thomson as a young man by W. H. Townsend.*
*(Courtesy of the National Institute of Health, Bethesda, USA)*

The east side records his father and his two brothers who died in infancy:

*Set up in 1873 by*
*William Sharpey MD*
*in place of a stone then fallen in to decay*
*which bore the following inscription*
*erected by*
*Mary Balfour*
*In memory of her husband*
*Henry Sharpey*
*Who died on the 31st October 1801 in the*
*44th year of his age*
*and*
*Henry Sharpey*
*His son*
*Who died on the 19th December 1805 in the*
*7th year of his age*
*also*
*David Sharpey*
*her second son*
*who died 15th October 1808 aged 8 years*

Members of the Arrott family are buried in a separate grave; the stone records William Arrott, his wife Mary, who is given her maiden name of Balfour, not Sharpey, his four sons, William, David, Alexander and James, and his daughter Mary, but not Jacobina.

The grounds of Arbroath Abbey today are beautiful and tranquil; a site steeped in history and learning. What better place for the burial of a scholar for whom it was always his dearly loved home.

## HIS EARLY LIFE

It has been said that it was from the influence and teaching of his talented mother that he acquired his life-long love of learning, but of his early life we know very little. This is characteristic of all of the biographical sources which have been consulted; he was reticent about his private life and left little insight into his thoughts or emotions about people or about issues in religion or politics. But he was never dull; he is spoken of as being a good dinner table guest with a fund of anecdotes and he maintained warm relations with a wide range of people.

Sharpey went first to a local "dame" school and then to the Grammar School but there is no record of his education there. In November 1817, aged 15, he entered Edinburgh University taking in his first year classics from Professor Dunbar and natural philosophy (physics) from Professor Playfair. In his second year onwards he took anatomy and clinical medical subjects.

Edinburgh had a considerable reputation in the world of medicine, attracting many students from the rest of Britain and from Europe. Unlike the situation in England, the university was open to all regardless of religious persuasion and students were free to live where they liked without having the expense of colleges. The University was founded in 1582 but the College of Barber Surgeons had been incorporated as long ago as 1505. The College of Physicians was founded in 1681 and the first Infirmary was built in1733. All these institutions developed their own systems of teaching, their qualifications, and status of membership, sometimes in harmony with each other sometimes not. Other learned societies were founded such as the Royal Medical Society in 1779, which Sharpey soon joined, thus displaying his early commitment to medicine (the Society later made him an honorary Member). There was also the Philosophical Society and the Royal Society of Edinburgh, founded 1783. Thus within the confines of what was then a small city there existed a close-knit medical community, with strong academic links, which had a considerable influence in the medical world at large.

Medical education at that time was not confined to the University and the Royal Colleges; there were a number of private schools which offered better teaching, particularly in anatomy. In Sharpey's time the University professor of anatomy was Alexander Munro III (1773-1859) who had succeeded both his father and his grandfather in the post but by then he was exhibiting all the signs of a dynasty in decline. He was a prolix but inaccurate writer and his textbooks were hardly used. Like so many other students, Sharpey studied at the distinguished private school of John Barclay (1758-1826) a man who came late into medicine, taking his MD at Edinburgh when he was 38, and then becoming a full-time anatomy lecturer for over twenty five years. He collected a teaching museum, later presented to the Royal College of Surgeons, and wrote a number of valuable books. His staff included Robert Edmond Grant, (1793-1874), later to become one of Sharpey's colleagues in London, who was an assistant lecturer during Sharpey's early years. He was succeeded by Robert Knox (1793-1862) who took over the school when Barclay became too ill, during the last two years of his career. Knox was a brilliant lecturer who attracted the largest number of students ever to attend such classes in Edinburgh. Although never a teacher or a colleague of Sharpey, such was his reputation that Sharpey dedicated his MD Thesis to him in 1823.

The schools of anatomy throughout the country suffered from an acute shortage of bodies for dissection. Legally they could obtain only the bodies of executed criminals but some were imported from Ireland, hidden among other cargoes, and many were taken from newly dug graves by the so-called resurrectionists. The notorious case of Burke and Hare, who murdered in order to obtain bodies without the risks of grave robbing, culminated in the execution of Burke in January 1829 after Hare had turned King's evidence against his partner-in-crime. There was an outcry against Knox since one of the victims was delivered and sold to his dissecting room; a popular limerick of the day ran:

> Doon the close and up the stair
> But and ben wi Burke and Hare
> Burke the butcher
> Hare the thief
> And Knox the boy that buys the beef

Never a man to succumb to popular pressure easily, Knox was nevertheless forced to resign in 1837. After trying to rehabilitate himself as a lecturer in Edinburgh and Glasgow, without success, he ended up in London without the position or the prestige that his talents merited.

Sharpey knew Knox from his Edinburgh days and when they were both students in Paris in 1821 – 22; it has been suggested that the two might have contemplated going into partnership. However, when the scandal arose, Sharpey joined in the hue and cry against Knox and there is no reference to him in any of the letters to Allen Thomson even when Knox was living reasonably near-by in London. There might have been a question of divided loyalties since Sharpey's close friend Syme took over as Curator at the College of Surgeons from the ousted Knox who was no longer part of the anatomical establishment. Sharpey's failure to support his erstwhile colleague at this difficult time led to a long-standing coolness between him, on the one hand, and Grant and Wharton Jones (another London colleague who appears later) who had remained loyal to their fellow anatomist. Sharpey's half-brother James Arrott, who qualified in Edinburgh in 1827, was an assistant to Knox and may have been the unwitting agent in accepting bodies into the dissecting room. Despite the later obloquy, he retained a life-long regard for his former master.

A consequence of the grave robbing scandal (which, despite the notoriety of Burke and Hare, was not confined to Scotland) was the passing of the Anatomy Act in 1832 which made it legal for unclaimed bodies to be made available to anatomists by local authorities, a move which quickly put an end to the resurrectionist's trade.

These events took place after Sharpey had qualified in medicine in 1821 with the diploma of the Royal College of Surgeons of Edinburgh, the usual qualification for practice at that time, a university degree being necessary only for those with higher aspirations.

Immediately after qualifying Sharpey went to London to continue his studies. He spent some months at the private anatomy school of Joshua Brookes FRS (1761-1833) in Blenheim Street, a school with a high reputation for success in the examinations for the MRCS diploma. Brookes was a devoted scholar as well as a brilliant teacher; he poured his fees, which were lower than in other schools, into building up a museum of comparative anatomy which was second in size only to that of John Hunter. While studying in Paris in 1782, before setting up his school, he had investigated the use of alum as a means of preserving human bodies especially in the summer months when putrifaction often prevented any dissection taking place. This discovery was communicated to the Royal Society in 1784 and led to his election as a Fellow in 1819. So successful was he as a teacher that he aroused the envy of the Royal College of Surgeons of England, which ran its own school, and, in 1822, used its statutory power to forbid any examinations being held in summer. This led to the demise of the Brookes school and to the destitution of Brookes himself who was forced to sell off piecemeal his valuable museum; a situation which aroused the rage of the Lancet, ever ready to challenge authority. An interesting souvenir of Sharpey's first visit to London is the ticket issued to him by Brookes for the course of lectures and dissection in June 1821, the year before the ban commenced (front cover). The syllabus of the course (back cover) states that the dissecting room was open from five am, before the summer heat made conditions impossible.

From London Sharpey went to Paris, then the centre of excellence for higher medical training, attracting students from all over the world; he stayed there for nearly a year learning surgery from Dupuytren at the Hotel Dieu and from Lisfranc. While in Paris he made one of several lifetime friendships, this one with James Syme (1799-1870) the Edinburgh surgeon, with whom he kept in touch personally and professionally, as will be seen.

A personal incident, which shows Sharpey as a staunch friend, took place in 1832. It has no bearing upon Sharpey as a scientist or administrator but it does provide an example of the emotive issues which occur in every age and in every profession and which, in those days, were carried on with remarkable vehemence. A chance meeting in an Edinburgh street between Robert Liston (1794-1847), then an up-and-coming surgeon, and Fanny Willis, Syme's unmarried sister-in-law and sister of Robert Willis MD, (who appears later), led to Liston attending a party she was giving at the home of her friends Dr. and Mrs. John Mackintosh. This infuriated Syme to the extent that he forbade Fanny to have any more contact with the Mackintosh's, which she was to do so in writing, otherwise Syme and his wife Anne, Fanny's sister, would disown her.

Mackintosh (died 1837), eminent in the Edinburgh medical world, felt that his honour had been impugned so he published a pamphlet, which he sent to all his friends, castigating Syme in very forthright terms. Syme responded in kind and when neither would offer the first regret or apology, friends were asked to intervene. As a service to his friend Syme, Sharpey called upon Mackintosh to try to bring about peace or, strange as it may seem in that cultural Athens of the North, to make arrangements for a duel. Mackintosh found discretion the better part of valour and told Sharpey that although he would have given Syme the satisfaction of a gentleman, he was not now inclined "to place his body unnecessarily as a target upon which Mr. Syme may try to restore his lost honour". Duelling was still practised in Great Britain; only a few years earlier no less than the Duke of Wellington had challenged the Earl of Winchilsea. The circumstances have a slight bearing on Sharpey's life in so far as the quarrel arose over the Duke's support for Catholic emancipation; the Earl considered this to violate the Protestant constitution of the newly established King's College, London, of which both were patrons. He withdrew his financial support for the College and made public a letter in which he accused the Duke of dishonesty. A duel was arranged for Saturday the 21st March 1829, the duke's shot was wide and the Earl, realising his error, shot into the air and handed over a letter of apology. But opinion was increasingly against the combat and in law fatalities were treated as murder; as a consequence duelling fell into oblivion by the middle of the century. It is nevertheless intriguing to learn that the cautious and pacific Sharpey was very nearly called upon to prepare pistols for his dear friend.

The affair eventually petered out, no doubt to every ones satisfaction. The unfortunate Miss Willis had both her brother-in-law Syme and her good friend Mackintosh fighting for her honour and, as her letters show, she was truly perplexed. The origin of this storm-in-a-teacup was Syme's irascible nature and his jealousy of successful colleagues. He and Liston had worked together as teachers of anatomy but they quarrelled; Liston resented Syme's increasing success as a surgeon in a city where there were limited professional opportunities. Syme disapproved of Liston's acceptance, if not approval, of body-snatching to supply the dissecting room; and so their partnership ended in 1823. Later, Syme collaborated with Mackintosh in the Brown Square School of Medicine but the two fell out and Syme established his own School of Surgery. It was too much for Syme to have his sister-in-law on friendly terms with both his enemies and so he was led into a piece of gratuitously bad behaviour. However, it is pleasing to record that Syme renewed his friendship with Liston when the latter was appointed to University College London in 1834.

To return to the next stage in Sharpey's career, he returned to Scotland in 1823 and submitted a Thesis entitled *De Ventriculi Carcinomate* to obtain his MD degree from Edinburgh in August of that year. This was an important step to take if he was to progress in medicine beyond the confines of general practice and almost essential for an academic career which, in view of his studious nature, he must have been considering. To gain further experience, he returned to Paris for much of 1824 and then settled for a period in Arbroath where he assisted in his step-father's practice. It has been suggested that he was offered a practice of his own but he turned it down so as to avoid any competition with his family. However, his experience as a student and then as a practitioner led him to decide that he was more suited to the academic life and, with this in mind, he left home at the end of 1826 for an extended tour of continental medical schools. Several biographical notices refer to him setting out "with knapsack on his back and staff in his hand"; this is perhaps a romantic allusion to the idea of a poor but dedicated scholar overcoming penury and accepting hardship in the search for learning. Nothing is known about Sharpey's financial standing until he encountered the problems of retirement; he was never financially ambitious and he was certainly generous but, coming from a successful family, he is likely to have had the means sufficient for his needs as a student. The origin of this picture of him can be traced back to some biographical notes by his oldest friend Allen Thomson who, writing in old age, stated "Dr. S. made considerable parts of his foreign travels on foot with his knapsack on his back". This might be a pleasing fiction or a well-remembered recollection; it crops up again in connection with Robert Willis.

It is not possible to construct an accurate itinerary of his travels but he is believed to have travelled through France to Switzerland and then on to Rome and Naples. In the spring of 1828 he passed through Bologna on his way to Pavia (not Padua as sometimes stated) where he worked for a time with Bartolomeo Panizza (1785-1867) Professor of Anatomy. Panizza had studied cerebral localisation and identified the optic thalamus as the cortical centre for vision; he gave the first course on histology in Italy. From there Sharpey went via Venice to Austria and arrived in Berlin by August 1828 where he stayed for nine months undertaking a period of dissecting under Carl Rudolphi (1771-1832) Professor of Anatomy, an excellent teacher and investigator although against vivisection. He opposed the fanciful 'Naturphilosophie' then favoured in Germany; according to his famous pupil Johannes Muller (1801-1858) 'we owe much to his voice for having regained physicians from the field of medical miracles'. Sharpey returned to Britain via Heidelberg, where he worked with Friederich Tiedemann (1781-1861) Professor of Physiological Chemistry who had confirmed the existence of gastric acid and of protein digestion by the pancreas. He arrived back in Edinburgh later in 1829.

It was now his intention to become a professional teacher of anatomy in the extra-mural school. He had considerable experience of the continental schools and knew personally many of the leading figures. One further qualification was required, his admission as a teacher by the Royal College of Surgeons of Edinburgh. To obtain this he submitted a Thesis, called a probationary essay, entitled *On the Pathology and Treatment of False Joints*, for which he was admitted as a Fellow in 1830. This study of fractures which do not heal properly is considered later along with his other writings. Except for an unusual paper to be mentioned later, this was Sharpey's second and last contribution of a clinical nature, the first being his MD Thesis.

He visited Berlin again for three months but by the autumn of 1831 he was ready to start his career as a lecturer on anatomy in the Edinburgh extra-mural school with the advantages soon to be enjoyed from the passing of the Anatomy Act.

The extra-mural or extra-academical school did not refer to a single organisation but to several teaching establishments which were not part of the University. The College of Surgeons would be included as a school as well as a licensing body and there were also a number of completely private, fee-paying schools such as that of Dr. Barclay and the Argyle Square school which offered a full medical curriculum. According to Knox's biographer Henry Lonsdale, who taught at Argyle Square, "the reputation of Edinburgh as a school of medicine was mainly sustained at this time by the private schools clustered around the walls of the University. The teaching of the University had degenerated into a formalism, distasteful and repulsive to the student".

Sharpey went into partnership with Allen Thomson (1809-1894) who gave a course on physiology although subsequently becoming, like Sharpey, a full time anatomist. The professional association between the two moved to a close and enduring personal friendship which will be mentioned frequently and justifies a fuller biography of Thomson later in this chapter. Although much smaller than the other schools, Sharpey and Thomson were successful; the number of students increased in each of the five years of their partnership. This situation was not to last. The University, aware of its declining reputation, responded by bringing in a new regulation which made it compulsory to attend certain intra-mural courses as a prerequisite to the MD degree. As a result students could no longer afford fees for extra-mural lectures as well as at the University and the numbers at the former fell from 900 in 1826 to 326 in 1841. There were similar trends at the London extra-mural schools but by that time Sharpey was safely within the university fold.

Sharpey's school was situated at Surgeons Square where were located the old Surgeons Hall, dating from 1697, the Royal Medical Society, the anatomy school of Barclay and Knox, the Royal Infirmary and the old Surgical Hospital, formerly the High School. His lectures and dissections took place in a house at no. 9 Surgeons Square owned by Allen Thomson's father, John Thomson, Professor of Surgery in the University. It was a very satisfactory arrangement for Sharpey: he acquired teaching premises in the heart of the Edinburgh medical world, associated with a pillar of the University establishment and in partnership with a loyal colleague who became a devoted friend. It was a situation which could have persisted for the benefit of all concerned, but that was not to be.

After 1836, when Sharpey left for London, the house was acquired by the adjacent Royal Infirmary as a fever hospital but it was demolished in 1850 to make way for the new surgical hospital. In 1832 a new Surgeons Hall was built in what is now Nicolson Street and the old Hall was leased to Knox for his last years in Edinburgh.

Now settled, with an occupation and an income, Sharpey had time to turn to original work and he published a number of papers during the ensuing five years. These papers, which will be considered in detail with his later writings in chapter 4, brought recognition by the Royal Society of Edinburgh which elected him a Fellow in 1834. In retrospect, it was an important period of his life since at no other time did he undertake any original scientific work and it was partly due to the quality of what he did in this limited time that he obtained the professorship in London which led to his long period of influence on British physiology.

The demise of Sharpey's anatomy school took place very rapidly. In July 1836 he was approached by Richard Quain, a member of the recently established London University, now University College London, about a vacant position there as a Professor of Anatomy and he was appointed to it in August. There was just time to wind up his affairs in Edinburgh before the start of the next university session. How this appointment came about is the subject of the next chapter. A strong influence in this, and throughout Sharpey's life, was his friend Allen Thomson and it is now appropriate to provide a brief outline of the life of this distinguished Scottish academic.

## ALLEN THOMSON

For over fifty years there existed a mutually close and affectionate relationship between Sharpey and Thomson. There were many similarities in their careers and in their dedication to academic ideals and although for the most part Sharpey lived in London and Thomson in Glasgow they kept closely in touch by letter and with visits together whenever possible. Sharpey frequently stayed with the Thomson family on his vacations in Scotland and Thomson would attend meetings in London at the Royal Society or at the Physiological Society. The warmth and respect that they felt for each other is well illustrated in the letters in the Allen Thomson archive at Glasgow, a selection of which has been published by Jacyna in 1989. They are mainly from Sharpey to Thomson, 76 out of 99, and only 13 from Thomson to Sharpey; they address each other by surname throughout but whereas Sharpey signs formally as W. Sharpey, Thomson allows himself the near intimacy of being Allen Thomson. This convention apart, the letters are informal, sometimes even gossipy, and they give Sharpey the chance to "speak aloud", as it were, to an old and trusted friend on day-to-day matters in which they both had an interest. As will be seen, Thomson was influential, and generous in connection with Sharpey's move to London.

He was born in Edinburgh in 1809, the son of John Thomson (1765-1846) then Professor of Surgery at the Royal College of Surgeons of Edinburgh and subsequently Professor of both Military Surgery and Pathology at the University. His half-brother William Thomson (1802-1852) was also an anatomy teacher in Edinburgh for a time before becoming Professor of Physic in Glasgow.

He was named after his father's friend John Allen (1771-1843), a lecturer in physiology at Edinburgh. He attended the High School and the University and qualified MD and FRCS Edinburgh in 1830. He and Sharpey must have met earlier for there is a note recalling that they "became very intimate and constantly together" and went collecting pond life on the Braid hills in 1829. After graduating, Thomson went to study in Paris in the company of Robert Carswell (1793-1857) who had made anatomical drawing for Thomson's father and who was then on leave from University College; this was a family friendship of some significance for Sharpey, as will be seen. Thomson returned to Edinburgh in 1831 to join Sharpey, who had just completed his second European tour, in the new extra-mural school of anatomy. It was a successful collaboration with the number of student rising each year despite competition from the other schools. The partnership ended in 1836 when Sharpey accepted the post at University College and Thomson took an appointment as physician to the Duke of Bedford, based in London but often on travel abroad. He returned to academic life in 1839 when he was appointed to the Chair of Anatomy at Aberdeen. In 1842 he went back to Edinburgh as Professor of the Institutes of Medicine, a term used at that time to denote physiology, on the retirement of W. P. Alison (1790-1859). He stayed there until 1848 when he obtained the position which his father had always hoped his son would one day occupy, the Chair of Anatomy at Glasgow. He remained there for the rest of his working life retiring in 1877 after being deeply involved with many aspects of the university and with medical education. He moved to London to be with his son John Millar Thomson, then a lecturer in chemistry at King's College, and to pursue his many intellectual interests in the capital. He died there in 1884.

Like Sharpey, Thomson was essentially a dedicated teacher, not a research worker, but, unlike his friend, he did not generate a group of disciples who became the future leaders of the subject. Although not making any major discoveries, he was much respected as an expositor of his science. Like Sharpey, he contributed to Todd's *Cyclopaedia of Anatomy and Physiology* with a well-received article on Generation and he joined Sharpey as one of the co-authors of Quain's *Elements of Anatomy* through several of its editions. He also wrote a textbook *Outlines of Physiology* in 1848 based upon the course he gave in Edinburgh although by the time it was published he had become a full-time anatomist in Glasgow. Thomson was elected FRS in 1848 and he served as a vice-president of the Society. He became a member of the Physiological Society in 1878 and it was fitting that he was elected an honorary member of the Society in 1882 to fill the vacancy left by the death of his old friend Sharpey. Both men were also honoured by their former university by the conferment of the LLD degree, Sharpey in 1860 and Thomson in 1871.

This brief account does not do justice to the service he gave, as teacher and administrator, to the university of Glasgow but it will become clear that through his friendship with Sharpey he exerted an indirect but important influence upon physiology in England.

CHAPTER 2

# University College London

IN 1818, WHEN SHARPEY commenced his medical studies, there were four universities in Scotland: Edinburgh, Glasgow, Aberdeen and St. Andrews, whereas in England, with a much larger population, there were only two, Oxford and Cambridge, and entry to those was restricted to students professing the Anglican faith. When most of the capital cities of Europe had universities of distinction it seemed strange that London, with claims to be the capital city of a large part of the world, had none. In the 1820's a group of men, unattached to the social and political establishment of the day, brought out proposals to found a university in London which would be free of religious restrictions and non-residential thus making it available to those unable to meet the costs of college living at Oxford and Cambridge. The leading figures behind this were Thomas Campbell, the Scottish poet, and Lord Brougham, also of Scottish descent. Their proposals were made public in a letter to the Times on the 9th of February 1825 which was followed by a meeting on the 4th June at the Crown and Anchor tavern in Holborn at which Brougham presented his scheme to an audience of over a hundred. He was able to persuade enough members of the business and intellectual world to put up the necessary capital, thus becoming the shareholders, or proprietors as they were known, to purchase land, erect buildings and appoint staff. In October 1828 lectures started in The London University, as it was named, situated at the northern end of Gower Street. It did not have the Royal Charter which was necessary for the award of university degrees, being able to offer only certificates of completion of a course, and its applications to Parliament for these powers had been turned down.

The new university had not received universal approval, being opposed by the medical corporations, jealous of their rights to licence practitioners, by Oxford and Cambridge, which considered that their own degrees would be debased, and by those who deplored its lack of any religious recognition; the new college buildings did not include a chapel. In fact its constitution forbade the teaching of religion, priests of any persuasion were not allowed to serve on its council (nevertheless, the Reverend John Hoppus, a Baptist minister, was appointed, not without opposition, as the first professor of philosophy; the restriction in principle was eventually lifted). But it was felt by many that the metropolis should have a university institution which acknowledged the established faith, as enshrined in Crown and Parliament, and an alternative college was therefore founded with the Archbishop of Canterbury and other prelates on its Council. Support for this new project to rival "the godless college of Gower Street" came from the highest in the land including the Duke of Wellington, then Prime Minister, and King George IV after whom the new foundation was named. King's College was formally opened in 1831 and its regulations required that the day's work should commence with prayers in the chapel, hence it became known as "the godly college in the Strand". Unlike its secular rival, the new college had its own coat of arms based upon the Royal Standard with an angelic supporter on the left holding a crucifix (Sancte) and a bearded supporter on the right holding a book (Sapienter). On one occasion Sharpey, as a Scots dissenter, referred to this, mockingly, as the priest's arms.

From the outset the new College had a Faculty of Medicine of size and stature to compete with its rival at University College and it expected to have full university status with the right to award its own degrees. But the government did not like the idea of there being two competing bodies both offering degrees in the name of the University of London and it therefore set up a committee to look into the matter. The main recommendation of this committee was a masterly compromise. There was to be a single University of London which was an independent examining body into which would be incorporated the two autonomous institutions University College (now renamed) and King's College. These colleges, and any others which were recognised later, could teach and enter students for the examinations to be set by the parent body, the University, which alone could confer degrees. The University had its own Senate, Registrar and support-ing staff and was housed in a separate, government, building, originally rooms in Somerset House. The objections of Oxford and Cambridge and of the Royal Colleges were withdrawn and the new structure for the University of London was approved by an Act of Parliament in November 1836.

The composition of the original 38 members of the Senate showed a strong scientific and medical bias: 20 were Fellows of the Royal Society and 16 held medical qualifications, as did the first registrar, R.W. Rothman, and his successor, W. B. Carpenter. In comparison, there were only 6 lawyers and 2 bishops. Of their number 30 appeared in the Dictionary of National Biography. Sharpey was elected to this elite body in 1864 and served on it for the rest of his life.

One of the founding senators who must have been well known to Sharpey, and possibly a source of influence, was Neil Arnott MD, FRS, (1788-1874) who was born near Arbroath into a family as well-known as the Arrotts and the Balfours. The Lancet might well have fulminated against more Scots in the seats of power but his record of service to the University, to science and to society was exemplary. Sharpey certainly met him socially and was sponsored by him for election to the Royal Society.

Most of Arnott's career was in London in private medical practice but he retired early and spent much of his time giving lectures and writing pop-ular accounts of science as well as being involved with the University of London. He was out of the same mould as Sharpey, a Scot with a deep respect for learning and a passion to pass on his knowledge to younger gen-erations; he bequeathed the Neil Arnott Scholarship in physics and chem-istry to the University and he also made generous grants to Bedford College and to the Scottish universities. He was the brother of James Arnott (1797 – 1883), also a doctor, who did much to promote the use of cold as a local anaesthetic.

It was envisaged from the start that the new London University (hence-forth referred to as University College to avoid confusing it with its later parent body, the University of London) should have a medical faculty which would confer academic status on what had been, in England, mainly a craft based upon apprenticeship and examination by the medical corporations.

In England, only graduates, usually from Oxford and Cambridge, were eligible for Fellowship of the elite Royal College of Physicians which entitled them to practice in London and the surrounding area. The majority of students at the new university, although now able to take a degree, opted for the diploma qualifications from the Royal College of Surgeons (MRCS) and Society of Apothecaries (LSA) after completing their course of medical studies. This was a quicker and more certain route than working for a degree which, in the early days of the university, was not itself a licence to practice and still had to be followed by the above diplomas.

The Medical Faculty of the new university was beset by a number of crises almost from its inception. There was a dispute over the location of clinical work which, it was wrongly assumed, would be done at the nearby Middlesex Hospital since that was the clinical home of Sir Charles Bell (1774-1842), the prestigious first Professor of Surgery and Physiology. The Middlesex had its own, independent, board of governors who were not consulted about the proposed arrangement. They feared that an association with the "godless college" might discourage financial endowments for their hospital and they withdrew their support for the joint venture. Faced with pre-clinical students who had no places for clinical work University College decided to found its own hospital and by 1835 the new University College Hospital had been built on the opposite side of Gower Street. One consequence of this was the creation of a separate medical school at the Middlesex to provide its own intake of students to the hospital.

There were a number of well-publicised personal quarrels at University College, which led to the resignation of Sir Charles Bell in 1830, and the turmoil involving staff and students which resulted in the dismissal of Granville Sharp Pattison as Professor of Anatomy in 1831. He was replaced by Jones Quain, a graduate of Trinity College, Dublin, in classics and in medicine, who taught anatomy in the Aldersgate school in London where he commenced writing his famous *Elements of Anatomy*. He was an accomplished scholar with interests in literature and of a mild disposition which was not in harmony with the sometimes turbulent atmosphere of the medical school. He was assisted, as a demonstrator, by his brother Richard Quain who was a difficult colleague (but an important figure in Sharpey's life). The brothers did not get on well and finally Jones Quain decided to resign and devote himself to a less stressful life; he lived in Paris for many years and made no further contribution to the University.

His letter of resignation was written from his home in Cork on the 16th June 1836 and accepted by the Council on the 16th July with expressions of regret and of gratitude for his invaluable service to the University.

Appointments to the academic staff were formally the responsibility of the Council but there was a recognised process of consultation to be followed. Firstly a committee of five from the Medical Faculty of the Senate had to consider whether to seek candidates privately or to advertise publicly and to advise Council accordingly. Its members attended this Council meeting to deliver their recommendation and it was agreed to advertise with the final date for applications being the 3rd August.

The Senate met on the 23rd July when the Dean of the Medical Faculty presented a four point plan for filling the vacant post. In essence, this was to appoint two professors of equal rank and to allocate the syllabus between them; one of the professors was to be Richard Quain, the other yet to be chosen. This plan was put before Council at its meeting on the 30th July but consideration was deferred until after the applications had been received.

The Council met again on the 3rd August; having received thirteen applications it was able to send them to the Senate for detailed consideration. Senate next day delegated this task to an appointments sub-committee of five, chosen by ballot, which submitted its report to the Senate on the 6th August. This body then voted on the recommendations and sent them on to the Council for its meeting on the 6th August. The Senate on that occasion consisted of nine professors who voted by six to two with one abstention to accept their subcommittee's nomination of Sharpey. The report was very detailed, making comments on all the candidates in turn, and going through it took up most of the Council's time so that a final decision had to be deferred until its next meeting on the 11th August. Sharpey was then elected Professor by nine votes to two and he was introduced to the Senate on the 13th August. He was aged 34 when he took this decisive step in his career.

The appointment of Sharpey as Professor of Anatomy and Physiology was not as straightforward as this account may suggest, not surprising in view of the College's propensity for turbulence. In this apparently routine piece of university business Richard Quain may be seen to have played a somewhat devious and self-interested part but, it must be said, he never harmed Sharpey's interests then or in the future. The closeness of his association with Sharpey for over forty years makes him a figure of some importance in this narrative and an account of his life is therefore of interest.

## RICHARD QUAIN

There were four members of the Quain family at University College and since three of them were named Richard they are often confused with each other. Jones Quain (1796-1865) and Richard Quain (1800-1887), the anatomists, were brothers, the sons of Richard Quain of Fermoy, County Cork, Ireland, and his first wife, formerly a Miss Jones. After her death Richard Quain senior married again and had a son John Richard Quain (1816-1876) who qualified in law at University College, took silk and became a notable judge for which he received a knighthood. He died unmarried and left his fortune to his half-brother Richard, above.

Richard Quain senior had a brother John whose son Richard (1816-1898), the cousin of the above Jones and Richard Quain, became Sir Richard Quain FRS, physician to Disraeli and Queen Victoria and author of the famous *Dictionary of Medicine*. He had been a pupil of Sharpey, winning a gold medal for physiology and graduating MB in 1840.

Richard Quain the anatomist, and colleague of Sharpey, was born in Ireland and apprenticed to a local practitioner before going to London to help his brother Jones at the Aldersgate School. As was then usual, he spent some time studying in Paris where he attended the lectures of Richard Bennett, an Irish friend of his father, who in 1828 was appointed Demonstrator in Anatomy at the new London University medical school. Quain joined him there as an assistant and was promoted to Senior Demonstrator when Bennett died in 1830. He succeeded Charles Bell as Professor of Descriptive Anatomy in 1832 and in 1836, on Sharpey's appointment, he became Professor of Anatomy until he resigned in 1850.

He had qualified MRCS in 1828 and in 1834 he obtained the post of assistant surgeon at University College Hospital which he held in addition to his teaching post in anatomy. On the re-organisation of the College of Surgeons in 1843 he became one of the newly created Fellows and was elected FRS in 1844. On the death of Liston in 1848, he was, "after a stormy progress", promoted to full surgeon and Professor of Clinical Surgery which he held until his retirement in 1866.

He held a number of well-recognised offices in the medical establishment: President of the Royal College of Surgeons in 1868, Hunterian Orator at the College in 1869, served on the General Medical Council 1870-1876 and was Surgeon-Extraordinary to Queen Victoria.

In 1859 he married Viscountess Midleton, a widow, but she died before him and there were no children. He bequeathed his fortune of £70,000, inherited in part from his half-brother, to University College which founded in his honour the Quain Professorship of English Language, Quain Studentships in English and Biology and a Quain English Essay Prize. Earlier he had given his medical books and his half-brother's law books to the College library. He edited his brother Jones's *Elements of Anatomy* through several editions and was the author of a superbly illustrated work *The Anatomy of the Arteries of the Human Body* (1844). He also published *Diseases of the Rectum* (1854) and *Clinical Lectures* (1884). He took a keen interest in medical education, writing a number of pamphlets on the subject and stirring the Royal College of Surgeons to make changes in their examination requirements.

This exemplary career, marked by his devotion to scholarship and to University College, was marred by personal qualities which could not be ignored even by his most loyal obituarists. According to them, he was a short and extremely pompous little man who went round his wards with a slow and deliberate step and always wearing his hat. He was an unamiable colleague with a jealous nature ever ready to impute improper motives to those who disagreed with him. He is reputed not to have spoken to Erichsen for fifteen years after the latter was favoured for the chair of surgery. He was said to have quarrelled at one time or another with most of the staff of University College Hospital. He was, nevertheless, a man of liberal principles and independent views who was quick to defend those who he thought had been unjustly attacked and the popularity of his lectures often exceeded the capacity of the lecture room.

Quain defended Sharpey's interests from the time of his first appointment in 1836 up to obtaining a retirement pension for him in 1874. In turn,

Sharpey never criticised Quain in his usually frank letters to Allen Thomson. The two had a mutual respect for each other and although they were never on terms of personal friendship they remained loyal colleagues.

## SHARPEY'S APPOINTMENT TO UNIVERSITY COLLEGE

On the 16th July, the same day as the Council commenced proceedings, Quain wrote to Sharpey urging him to send in an application. His letter reads:

> In much haste for post
>
> My dear Sir,
>
> I spoke to you of a vacancy in our School of Medicine occasioned by the resignation of Dr. Quain - If you continue to think an appointment with us is desirable, I suggest the prudence of your making an application within a week or so. I am strongly of opinion it would be to your advantage to come to town within a week or ten days or a little more- the sooner the better. The exact nature of the office, the probable emoluments and all particulars could best be known by your personal application here - you may get a substitute for your Demonstrations or give them twice a day - or return to finish them after a short delay.
>
> I am Dear Sir
>
> Very truly yours
>
> R. Quain                    do not omit testimonials etc.
>                                           If to be easily had.
>
> 23 Keppel St.
>
> London
>
> Saturday 16 July

This letter, which still exists at University College, is a very informal fragment, a single sheet folded to make a note about five by three inches in size, closed with a red wax seal and addressed to "Doctor Sharpey, (Lecturer in Anatomy), Edinburgh" with a charge of half a penny imposed (this was before the introduction of adhesive postage stamps in 1840). Quain had written on the outside "Excuse the address, I know not your residence" (Sharpey was then living at 3 Alva Street but the letter might have been delivered to his school in Surgeons Square).

Sharpey received this letter on Monday the 18th July. On the previous day (Sunday deliveries being then in operation) he had received a letter from Allen Thomson, written from London:

> 15th July 1836
>
> My dear Sharpey,
>
> The day before yesterday I got a summons from Carswell to see him in some way or other as soon as possible and when I called on him I found you were the subject of conversation. From what he said I think there is every likelihood of your being chosen by the Council of the London University, but of course we can have but a very imperfect notion at present. The Faculty of Medical professors were to meet today to discuss the matter & Carswell says that their suggestion is very well attended to by the Council. Mayo seemed to have an inclination to offer himself and Grainger is I suppose a candidate but Carswell seems to think that others as well as he is himself will be for having you.
>
> I need not say what a tumult of feelings al this raised in my mind.
>
> The emoluments of the situation appear to be nearly £800 a year and you will deliver your lectures in much more advantageous circumstances there than you can do in Edinburgh. It increases your chances of the Edinburgh chair and of preferment in every way and as far as I can see is most advantageous for you.
>
> I wrote you thus promptly in case it were possible for me to be of any use to you in this matter. Write me at all events and let me know what you are thinking of the matter.
>
> I weary to hear something of Surgeons Square. Indeed I feel melancholy whenever I think of it. & what it will be when you are away also?
>
> Yours ever,
>
> > Allen Thomson
>
> Address to me care of John Murray Esq., 50 Albermarle Street if you don't put your letter under cover to the Duke (of Bedford).

Having already received Quain's letter, Sharpey replied at once to Thomson:

Edinburgh 18th July 1836

My Dear Thomson,

Many thanks for your kind letter. I am quite alive to the nature of its contents, but I assure you that when I received it you were better aware of what was going on than I was. Mr.R. Quain heard a lecture from each of the Teachers here without their knowing of it, he called on me after and spoke about the change about to take place asking me at the same time if I would be likely to come forward as a candidate. I informed him that I would relish teaching the branches he mentioned in a chair of the Lond. University very greatly indeed. I met him afterwards at dinner but no more passed between us on the subject, and your letter which I got yesterday was the next intelligence I had. This afternoon I had one from Mr. Quain advising me to come up to London as early as convenient. I called on your brother, and the result is that I start for London on Saturday. I can easily do this as the Session ends next week.

Dr. W. [William Thomson] bids me request you to go over the list of Council with Mr. James Mylne- lest he know any one he could come at, he also spoke of Mr. J. Murray knowing Mr. Greenough who is one of the Council. I fear many of them may be inclined to listen to Somerville and there is one whom I certainly could not expect to look favourably on me. I need not say I mean W—n. [Warburton]

"I would naturally say a great deal more to you my Dear T. on this matter, and what plans in the event of success we might pursue, but I feel that the prospect is still too distant to permit me (naturally the reverse of sanguine in my disposition) to indulge in what might turn out to be day dreams".

He then continued to write about some physiological experiments he was conducting and of papers to appear in Todd's *Cyclopaedia of Anatomy and Physiology* and concludes:

"I would say more but till the London affair is over, one way or another, I shall feel in an uncomfortable state.

I am my dear Thomson

Yours most sincerely

W. Sharpey"

Sharpey sent in his application on the 29th July; it was a brief, almost a terse, message stating "In offering myself as a candidate for the office of Professor of Anatomy in the University of London I beg to state the following particulars respecting my education and subsequent pursuits". There followed a *curriculum vitae* of only six lines in which he mentioned his year in Berlin "in the special pursuit of anatomical and physiological knowledge", his only allusion to the fact that the professorship was in physiology as well as anatomy. He gave some details of his anatomical teaching and a list of several well-known Edinburgh men who would provide testimonials (Sir George Ballingall, professor of military surgery; W. P. Allison, professor of physiology; R. Christison, toxicology; D. Craigie, histologist and editor; John Thomson, professor of general pathology; R. Jameson, mineralogy; D. Maclagan, former President of the Royal College of Surgeons of Edinburgh; James Syme, professor of surgery; J. Abercrombie, physician)

In answer to a request, he wrote again providing more details of the size of his classes since he opened his anatomy school:

| Session | 1831-32 | 1832-33 | 1833-34 | 1834-35 | 1835-36 |
|---|---|---|---|---|---|
| Number of students at lectures | 22 | 53 | 62 | 72 | 71 |
| Number of students at practical | nil | 39 | 68 | 87 | 88 |
| Number of new students | | 14 | 38 | 43 | 51 |
| Total number of students | 22 | 67 | 100 | 115 | 122 |

He mentioned that although his was the smallest school in Edinburgh, "my class has been entirely made by myself" compared with that of Dr Knox, who inherited John Barclay's class, and of Alexander Lizars, who inherited the class from his brother John Lizars.

An historical point to note is that he did not offer practical dissection classes until his second session by which time the Anatomy Act of 1832 made the supply of bodies legal.

The outcome of his application was a happy one for him and he was able to write to Thomson:

(London 11th August 1836)

My dear Thomson,

I was not in time for the post.

I was elected today Professor of Anatomy, subject to such regulations as the Council may deem expedient.

The condition attached is merely this. I have been elected to fill the vacancy occasioned by Dr. Q's resignation, in short into his place, but the contemplated and wished for alterations of duties etc. have not yet been adopted, it was felt that this would delay the matter and the present proceedings makes all right. It was fortunate that the discussion respecting the allotment of the duties was not entered on, as time would have been lost and perhaps another adjournment been the consequence.

There was a division in the council but this is private. I had an immense majority. 9 to 2. Don't mention this.

My great desire is now to do justice to the appointment.

I will leave this Saturday. I think by the Dundee steamer. I will be a couple of days in Arbroath and see you in Edin. on Thursday evening.

This is my present intention but I may change it.

Yours most sincerely

W. Sharpey

There was obviously much lobbying going on in favour of Sharpey but entirely without his knowledge or consent. When Thomson wrote to him it was clear that Sharpey was to be the choice of Quain, Carswell and possibly others, if he could be persuaded to apply, even though at the date of the Carswell - Thomson meeting Council had not yet decided to advertise the post. And even earlier, before Thomson's letter, Quain had visited Edinburgh, heard Sharpey lecturing and then made contact with him to see whether he was interested in applying. It is not known exactly when this visit took place. It must have been in the three weeks between the receipt of Jones Quain's resignation, about Monday 20th June, and Carswell's letter to Thomson, in favour of Sharpey, on Tuesday 12th July. Since the journey to and from Edinburgh, by road or sea, would have taken at least two days in each direction, there was little time for Quain to visit all the medical schools and return to London in time for meetings of the Medical Faculty, Senate and Council. The visit is referred to in Quain's obituary: "the fame of Dr. Sharpey having reached London, Mr. Quain paid a visit to the northern capital and attended Dr. Sharpey's lectures *in cognito* for the purpose of ascertaining whether the reports of his excellence were correct". Thomson also mentioned the inspection in his memoir, writing: "I happened to be in the classroom at the time and remember very well his coming in and listening to the lecture without any warning to Dr. S."

The Edinburgh correspondent of the Lancet writing on the 28th July, before the closing date for applications, found the whole situation very gratifying, being yet another example of the superior talents to be found in the north. He wrote:

> "No sooner had intelligence reached us of the resignation of Dr. Jones Quain, than the most varied speculations arose about the vacant chair, and every iron was put in the fire by our worthies here to procure the seat for some friend. They argued that from the liberal supply of luminaries which had already been exported from modern Athens, to fill the chairs of your schools, a fair chance existed for more employment for us on the present occasion. Guess, then, our feelings of surprise and delight, on the arrival, certainly very unexpected, of one of the professors of your eminent institution [ie. Quain]. The learned professor was received with open arms, and not a doubt was entertained but that one or other of us would be in London before a week could elapse. It is whispered here that your boasted system of concours consists of nothing more than the most interested of your professors coming amongst us as a great unknown, quietly attending the lectures of the different teachers, and selecting the one who he thinks is the most likely to accommodate himself to the state of things in your quarter".

Edinburgh gossip was well-informed since the correspondent went on to add "It is right that it should be known in your city that the selected gentleman is one of the chosen offsprings [ie. Sharpey] of Dr. John Thomson whose name is already so illustrious in the annals of chair-filling". However, this issue of the Lancet was not published until the 13th August by which time the Senate had selected Sharpey, unaware that its decision had been anticipated.

Quain was obviously satisfied that Sharpey was the right man for the post, a view no doubt supported by their dinner time conversation since Sharpey was eminently sociable on such occasions.

Quain must have reported back to his friends and then left it to Carswell to add additional pressure on Sharpey through the good offices of Allen Thomson who wrote to his friend so enthusiastically. Not only did he personally urge him to apply but he also sent, without Sharpey's knowledge, a fulsome letter of recommendation to Carswell which was later referred to by the Senate committee.

It is worth examining the composition of the various bodies which were involved with the appointment of the new professor to see whether any characteristics of the members can be identified which might have been expected to favour Sharpey (Note 2).

The Medical Faculty Committee, which sought an outside appointment to the chair by means of a public advertisement, consisted of Carswell, Davis, Grant, Quain and Thomson. There was a very strong Scottish background; three of them were born in Scotland, four, including the Welshman, Davis, had qualified in Scotland and the fifth, Quain, was Irish; altogether a very Celtic influence on an English university. Two of them, Carswell and Grant, could be described as being on friendly terms with Sharpey and they would certainly have known of his reputation. They may have spoken to Quain about him or Quain himself may have taken the initiative to make the exploratory journey to Edinburgh. There is no suggestion that any of them, other than Quain, stood to gain anything by Sharpey's appointment, unless it was to block someone less congenial as a colleague. They all held chairs which they retained for the rest of their careers at University College and personal ambition was unlikely to have been an important factor.

The Senate Committee, which considered the applications, consisted of Malden, Carswell, Elliotson, Thomson, and Turner. Thus to the Scottish connections of Carswell and Thomson there could be added Elliotson and Turner, who were both educated at Edinburgh, leaving only Malden as an outsider, in this context. All five members of this committee were strongly academic in outlook, settled in their careers and able to seek the best to maintain the highest standards in the new university. Their unanimous choice of Sharpey was entirely consistent with their background leaving aside any influence which Quain may have exerted.

The Senate was divided between the Faculty of Arts and the Faculty of Medicine with some members serving in both divisions. It was their duty to vote upon the report of their appointments committee before it was sent to Council. Those present at its meeting on the 6th August were: Booth,

Carswell, Cooper, Davis, Elliotson, Grant, Malden, Ritchie and Turner. As mentioned, they voted by six to two, with one abstention, to accept the nomination of Sharpey. Grant, Liston and Ritchie had sent a joint letter to the Council supporting Grainger but on the day of the Senate meeting Liston was absent so presumably Grant and Ritchie were the two who voted against. Probably Malden, as the only representative of the Faculty of Arts, preferred to remain neutral on the issue and therefore abstained.

The Council consisted of twenty four members but only eleven attended the final meeting at which the vote was taken. These were Waymouth, as chairman, Booth, Boott, Duckworth, Leader, Prevost, Romilly, Tooke, Tulk, Warburton and Wood.

As far as can be ascertained, there was no dominant medical, scientific or Scottish influence in the Council. Cambridge was well represented but this could not be seen as favouring Sharpey; only one member was medically qualified and the two Fellows of the Royal Society were not scientists. The law and Parliament were prominent and there was a wide range of outside experience to be tapped. Two of the above voted against Sharpey; one would have been Warburton, who Sharpey had already seen as an opponent, but the identity of the other is unknown.

It is in the nature of university councils to leave purely academic matters in the hands of the senate; the professors are in the best position to judge the quality of the applicants and to decide who they want as a colleague, a view already expressed by Thomson in his letter to Sharpey. This was the case here and Sharpey was elected by a large majority.


THE CANDIDATES

The Senate Committee report, on which both Senate and Council made their decision, contained the names of thirteen applicants (Lucas, Pettigrew, King, Hiller, Solly, Hunter, Mayo, Lizars, Grainger, Hart, Grant, Sharpey and Quain).

Most of them were dismissed on the grounds of too little experience as teachers and not enough knowledge of physiology. The latter was rather surprising given the state of physiology in 1836 and the lack of any experience by the Committee of what little physiology then existed. It is hard to avoid the conclusion that Sharpey was the favourite for the post from the start and that interviewing the others was a formality. Lucas and Pettigrew were dismissed in a few words.

Lucas, probably P. Bennett Lucas MRCS 1833, was a lecturer in the Hunterian Medical School. Pettigrew, has not been identified. He could have been Thomas Joseph Pettigrew, a rising star in the more scholarly aspects of medicine and already an FRS. Thomas Wilkinson King MRCS 1830, a lecturer in pathology at Guys Hospital, was criticised for lack of judgement in submitting a testimonial from Magendie which related only to a course he had attended when a student in Paris many years earlier. George Leicester Hiller MRCS 1829, a surgeon, had his application disposed of very quickly. Samuel Solly MRCS 1828, was a lecturer in surgery and physiology at St. Thomas's Hospital where he went on to have a distinguished career. Soon after applying for the University College post he was made an FRS for his work on the brain. He also had a strong interest in the arts and he might have made an excellent professor, but the Committee thought otherwise.

Robert Hunter, a Glasgow graduate who qualified in 1822 and had been a professor of anatomy there before moving to Westminster Hospital Medical School where he taught both anatomy and physiology. He was obviously a strong candidate but he, too, was rejected.

Herbert Mayo was dismissed very briefly, almost peremptorily in view of his academic status in London. He had studied under Bell at the Middlesex Hospital, took his MD at Leiden and in 1830 was the founding Professor of Anatomy and Physiology at King's College. He remained there until 1836 when "his ill-judged and unsuccessful candidature for a vacant professorship at University College necessitated his withdrawal from King's". He then became surgeon to the Middlesex Hospital and (almost in revenge?) founded the medical school there in competition with nearby University College. He had become an FRS in 1828, the only other candidate with this distinction, and he had written a number of works including *Anatomical and Physiological Commentaries* (1822), which has a Garrison & Morton citation, and *Human Physiology* (1827). The latter book was severely criticised in the Lancet, which questioned his integrity, and there had been a dispute with his former teacher, Bell, over the question of priorities surrounding the Bell-Magendie law. On the other hand his biographer speaks of him as a cultivated man who was much respected by his students. It is superficial to suggest that rivalry between University College and King's, much loved by later generations of students, played any part in the Senate's decision but their report ignored all his contributions and stated only that despite the great opportunities he had at King's "he has not been very successful a lecturer and Professor". The brevity hints at personal dislike or a hint of scandal. The Lancet also opposed his appointment in its characteristically outspoken way: "His failure in the Strand furnishes his chief claim to the consideration of the Council. If the Council are fond of empty benches and empty coffers let them by all means choose Mr Mayo" and later: "Mr Mayo, by offering himself for the vacant chair in the University of London, does not appear to be more anxious to quit the temple of bigotry in the Strand, than are his colleagues desirous that his feelings in that respect should be promptly gratified".

Alexander Lizars, was running the private medical school in Edinburgh founded by his brother John but he did not supply any testimonials and the committee found it impossible to judge "his knowledge of physiology, his habits of research or his manner of lecturing"; possibly Quain heard him in Edinburgh and passed on a poor opinion of him.

Richard Dugard Grainger (1801-1865) was a lecturer at the Webb Street school which he had inherited from his brother in 1824 and continued to run very successfully with over 200 pupils. This did not satisfy the Senate Committee members who complained that since the school was so close to Guy's and St. Thomas's hospitals it had the geographical advantage of an excellent catchment area, an advantage made better (or worse, depending on the view taken!) by the fact that it charged much lower fees than the hospitals. It was admitted that it had a better pass rate than its neighbours in the examinations of the Society of Apothecaries and the Royal College of Surgeons but, even so, it was not as good as University College. This showed that "he is below the standard to which the University of London has a right to look for". His textbook *The Elements of General Anatomy* was dismissed as being merely a compilation which had not been reprinted. It seemed as though he couldn't win!

But worse was to come. "They had no evidence that Mr. Grainger is a gentleman with a cultivated mind and extensive general knowledge. He had not had the advantage of a regular Academic education. It was from this defect …that he had been passed over in favour of Jones Quain in 1831. Finally, the *coup de grace*, "The Committee are informed by Dr.Davis upon his personal knowledge that Mr. Grainger's personal intonation in lecturing is, in is opinion, peculiarly unpleasant".

Grainger was a good candidate who deserved better treatment than this. After the closure of the Webb Street school in 1842 he became a lecturer at St. Thomas's where he was much liked by his students. They offered him a generous gratuity on his retirement (£500) which he turned down in favour of providing a prize in physiology. In 1848 he gave the Hunterian lecture and he was involved with many medical charities and other good works. His book *Observations on the Structure and Function of the Spinal Cord* (1837) supported the work of Marshall Hall on reflexes and advanced the understanding of the sympathetic nervous system. His attitude to physiology was in the modern, mechanistic mould which denied the existence of unique vital processes not subject to the laws of chemistry and physics. In one published lecture he cited the work of Prout, Liebig, Dutrochet and Magendie as examples of the new objectivity of physiology and he approved of vivisection when carefully conducted. In 1836 he was no less physiologically inclined than Sharpey and he might well have made a successful professor, but if you lecture with a Birmingham accent keep away from Gower Street!

It has been noted that three members of the Senate (Grant, Liston and Ritchie), who were not on the Committee, wrote a letter of support for Grainger declaring that he "is much more likely to advance the interests of the University …than the appointment of a teacher from a distance whose merits & success are comparatively unknown", a view of Sharpey later taken up by The Lancet. A last-ditch attempt to promote the claim of Grainger was made by Warburton at the next meeting of the Council when he read a letter from Grainger written the previous day detailing the progress of his professional and scientific education. This evoked no comment. Grainger had obviously been made aware (by Warburton?) of the remarks about him and was hoping to improve his standing but it was to no avail.

Hart was a lecturer on anatomy and physiology at the Park Street Medical School, Dublin where he attracted large classes. He had glowing testimonials but the Committee found him weak in his knowledge of foreign work (one of Sharpey's strengths) and so rejected him.

Robert Edmond Grant (1793-1874) was already the Professor of Comparative Anatomy in the University but his classes were small and his income was accordingly very meagre, of the order of only £150 a year. It would have been to his advantage to move to a position where a large number of medical students were involved but for some unaccountable reason he had made it clear that he was only interested in the post if it were restricted to physiology. This was not going to happen and so his application was not taken any further. Despite his voluntary withdrawal from the race, The Lancet considered him to be the best candidate but this was mainly on the grounds that it favoured anyone other than Sharpey, who long remained its *bete noir*. Grant, already an FRS, was undoubtedly a talented scholar with more publications to his credit than Sharpey but he remained outside the medical world despite being qualified, and he was not of a sociable nature so his application was unlikely to impress. He and Sharpey remained on good terms as colleagues and Sharpey wrote a warm obituary of him for the Royal Society.

Sharpey's application received a very laudatory response, as the following extract from the Council Minutes for the 6th August shows:

> "Of all the candidates for the vacant chair of Anatomy and Physiology, the Committee are unanimously and decidedly of opinion, that the best qualified is Dr. Sharpey.
>
> Dr. Sharpey became a Licentiate of the College of Surgeons in Edinburgh in 1821 and afterwards studied for 3 months in London, and a year in Paris. He took his degree as M.D. Edin. in 1823, then went again for some months to Paris, and afterwards was for some time engaged in practice assisting a relation. In 1827 he again visited the Continent, and proceeded to Italy and Germany & spent a year at Berlin. The express object of his travels was to inspect the great medical institutions of various countries; to become acquainted with the researches of foreign scientific men, and to study the methods of medical education in different schools
>
> In 1830 he became fellow of the College of Surgeons in Edin. He began to teach anatomy in 1831, and since that time has been employed in that manner. In the first session in which Dr. Sharpey lectured his class consisted of 22 pupils. In the last session his anatomical lectures were attended by 71 pupils; his demonstrations by 89; and the total number of individual pupils was 122. This number is rather larger than the number assigned to Dr. Sharpey in the extract from the album of the College of

Surgeons furnished by Dr. Lizars. Dr. Sharpey has explained that this difference arises from some pupils neglecting to register their tickets. Of course a proportionate addition must be allowed to the number of Dr. Lizars. In comparing the attendance upon these two classes, it must be observed, that Dr. Sharpey's is a new class formed by himself, but that Dr. Lizars succeeded to a large class formed by his brother; & that Dr. Sharpey's is a class of Anatomy & Physiology conducted by himself and one colleague, while the class of Dr. Lizars is a portion of the established medical school in Argyle Square, which has a full array of teachers and supplies the means of a complete education to pupils.

Dr. Sharpey has sent his publications on points of Pathological Anatomy, Comparative Anatomy & Zoology. The most elaborate is the article on "Cilia" in the *Cyclopaedia of Anatomy and Physiology* which Dr. Bostock has read and pronounced it to be "an excellent performance indicating a correct judgement and well furnished mind". Dr. Elliotson has examined likewise Dr. Sharpey's account of Professor Ehrenburg's researches on Infusoria and approved highly of it.

The testimonials in recommendation of Dr. Sharpey are exceedingly strong, and proceed from gentlemen of the highest eminence in the practice of the medical profession in Edinburgh, and from the most distinguished Professors in the University - Dr. Thomson, Dr. Christison etc.

One of the most interesting, which indicates the high attainments of Dr. Sharpey and the opinion & expectations formed of him by his medical friends at a very early period, is from Dr. Hodgkinson [ie Hodgkin] of Guy's Hospital who was his fellow student in Paris.

Upon a careful comparison of all the testimonials, it appears that Dr. Sharpey has made anatomy his peculiar study, and professes an extensive, very accurate, and exact knowledge of it in all its branches. To physiology also he has paid special attention, and altho less his professional subject as a lecturer than anatomy, it appears to be his favourite pursuit. His knowledge of physiology is extensive and accurate; and his physiological writings are held in high estimation. He has a good knowledge

of Comparative Anatomy. To these qualifications he adds considerable practical skill in the medical Profession. He is well
and thoroughly acquainted with the literature of his science and
profession, not only in his own language, but in French and
German.

His friends speak in very strong terms of his high talents,
his great zeal, and his unwearied industry in the continual prosecution of his studies. The point on which they almost all dwell
is the singular accuracy and minuteness and soundness of his
very extensive knowledge. He is an accurate observer, an original philosophical thinker, possessed of a very sound judgement
and cautious & logical in his deductions. He has had all the
advantages of an excellent education, and has very great general knowledge both of science and literature. His method of
communicating his knowledge both in speaking and writing is
described as plain, but peculiarly precise and clear, and consequently agreeable and attractive. The increase of his class is evidence of his success as a teacher. He is very attentive to his
pupils and has secured their respect and attachment. Professor
Jamieson testifies that Dr. Sharpey's pupils are marked in the
University for their enthusiasm in the prosecution of their studies. Dr. Carswell adds that it has been observed, that in the Theses
exhibited by them to the University, and in other essays, they
have shown original investigation & power of thought. Dr.
Sharpey has a great advantage as a teacher in having studied the
methods of instruction in the best continental schools. His moral
character is described as marked by probity and integrity,
straightforward, gentlemanly, and honourable. His temper and
manner are mild and amiable. He is of an obliging disposition;
and very generally respected & loved. Most of the letters which
are presented as testimonials, and which are not addressed to
Dr. Sharpey himself, bear the mark of strong personal esteem
and attachment. The Committee recommended to the particular notice of the Senate a letter of Dr. Alan [sic] Thomson the
present colleague of Dr. Sharpey as a lecturer. It is not unworthy of notice, and an indication of Dr. Sharpey's habits of mind,
that he is described as very punctual.

With these qualifications, established upon the strongest and
most trustworthy testimony, the Committee report to the Senate

their deliberate opinion that the claims of Dr. Sharpey to the vacant chair are very much stronger than those of any other candidate previously mentioned".

No other candidate received such detailed and glowing attention and it is clear that the majority of the committee had already decided who should be appointed. It would be hard for a Council with a very generalised outlook to find fault with such a panel of experts.

Quain was also a candidate at this stage and his application received careful consideration:

"with respect to his qualifications as a teacher they cannot express themselves in terms too high… they place him above all the candidates….In familiarity with general and structural anatomy and with physiology they believe him at least equal to any of the candidates except Dr. Sharpey. The committee are of opinion that the acquisition of Dr. Sharpey would be a very great gain to the University and they are exceedingly unwilling that it should loose the services of Mr. Quain….they recommend to the Senate that it should request the Council that it should mark their sense of Mr. Quain's eminent merits by according him the title of Professor of Anatomy and placing him upon a footing of equality with the new Professor of Anatomy and Physiology (as recommended by the Senate)".

Booth, as a member of both Senate and Council, then informed the Council that "at a meeting of the Senate just held the Report read had been adopted by the Senate on division 6 for, 2 against and that it should be presented to the Council as a report of the Senate".

At the resumed meeting of the Council on the 11th August Boott stated that he had authority for withdrawing the name of Quain from the list of candidates. Booth then proposed that Sharpey should be appointed, it was seconded by Took and passed by nine votes to two.

It will be noted that Carswell, in addition to sounding out Allen Thomson in support of Sharpey, was on three of the four bodies formally involved in the selection: the Medical Faculty Committee, the Senate Committee and the Senate. He was not a member of Council but, as he observed, that body seldom failed to accept Senate's advice on academic matters. Added to that, his friendship with the Thomson family must have made him a powerful advocate for Sharpey.

Nothing more was done about Quain's status until the 29th October when the Senate requested his formal appointment as Professor of Anatomy which was agreed by Council on the 5th November thus fulfilling one of his ambitions. He obviously viewed a clinical post as being his ultimate goal and in January 1837 he took a step in that direction when he obtained Senate approval for him to give lectures on surgery in Cooper's department. He went on to become a full surgeon and Professor of Clinical Surgery in 1848.

Quite why Quain took such trouble to obtain the appointment of Sharpey is not clear. He may have seen him to be the best of the possible candidates and in this respect he was acting solely in the interests of the University, but he made his first approaches before the closing date and so he could not have known with certainty who would apply. Perhaps he saw Sharpey as some one who would not be a competitor in the clinical field, his own ultimate goal. Grant would also come into this category but Quain, as his colleague, would have been able to judge him and in any event he withdrew from the race. It is possible that Quain had some prior knowledge of who had applied or who was likely to apply in the period between the 16th July, when it was decided to advertise, and the 3rd August. But he had started his own activities well before then, with his visit to Sharpey in Edinburgh, and Carswell was confident of Sharpey's success when he met Allen Thomson on the 13th July. The exact details may never be known but there is every indication of a well designed plan to bring about the appointment of Quain and Sharpey by somewhat machiavellian tactics, not unknown in the university world then, or now. The two certainly had cordial relations from the start when they both agreed about the allocation of lecture courses and fees which says much for Quain's notorious irascibility, which could be curbed, as it does for Sharpey's *sang froid*.

Quain's, and Carswell's, support for Sharpey seems to have influenced the Committee from the outset. He was not the only candidate with teaching experience and he had never taught physiology, unless this meant only microscopy. His research experience was less than that of Mayo (already an FRS) and he had not, by then, written any large scale works. He was not even the only Scotsman to apply, if precedent demanded that one of that ilk should join the fiefdom. Perhaps the points which weighed most in his favour were his evident devotion to scholarship (without thought to the fashionable world of London medicine), his wide knowledge of European science, his familiarity with the modern technique of microscopy, and his personal character in which the Committee saw an amiable colleague and an inspiring teacher.

For Sharpey, London would have had its attractions, perhaps more so following the death of his mother in May, with a consequent loosening of family ties. He had studied there after qualifying and would know of its strong cultural associations and he would have been aware that his Edinburgh anatomy school was small and therefore vulnerable now that the university schools were in the process of restoring their domination of medical education. As a full professor of a new and thriving university in the capital city he could have security and satisfaction and still have time to visit his beloved Scotland. It was a golden opportunity and he seized it.

The appointment of Sharpey did not please everyone. Public opposition was expressed by the Lancet in a series of editorials and letters which can now amuse us by the venom with which they were written. Even before Council had met, the Lancet was certain that " a series of intrigues have been in motion, in order to procure the appointment of his (Jones Quain's) successor". It regarded the Senate as a self-perpetuating body seeking to preserve its authority by appointing only those of a similar outlook and doing so by private subterfuge rather than by open competition to obtain the best. "In pertinaciously pursuing the system of secret self-election, the medical department of University College is in no respect superior to that of the least reputable of our medical monopolies". This view ignored the independent role of the largely non-medical Council although, as has been seen, Council tended to support the Senate on academic matters.

The Lancet pushed hard for the appointment of Grant, "a brilliant genius" and " a renowned member of the world of science", without substantiating this in any way, rather than "the unknown Dr. Sharpey of Edinburgh"; "where is the man who will enter into competition with Dr. Grant? Will Dr. Sharpey of Edinburgh? We smile at the thought". Warming to its theme it continued: "in this instance before us, we have hypocrisy, treachery, envy, and fraud superadded to the one ancient evil, *love of pelf*. Not only is monopoly rendered rampant once more, but the very triumph of the monopolists serves as an announcement that the community, the profession, and the students of the institution, HAVE BEEN BETRAYED. Who is Dr. Sharpey? Where is he known as a discoverer - as a physiologist? Who is Professor Grant? Where is that distinguished man not known as a discoverer and a physiologist? Yet Dr. Sharpey carries with him to the council a majority of votes of that body in favour of the UNKNOWN MAN OF THE NORTH!" But, bowing to the inevitable, it goes on to say: "We are willing to give Dr.Sharpey a perfectly fair trial, although we entertain insuperable objections to the system, under the sanctions of which he has been elected. It is against the system that we wage war, and not against Dr. Sharpey". Despite its thunder, the Lancet did publish some letters in support of the College as well as those which agreed with its views.

At his Start-of-the-Session address in October, Anthony Todd Thomson, Dean of the Faculty of Medicine, defended the appointment of Sharpey. He said "Some strictures on the mode of electing this gentleman have been published in one of the medical periodicals…the method adopted was that which is usual on such occasions…the Committee felt themselves fully justified in recommending him (Sharpey) as the fittest person to occupy the vacant chair…I went into the Committee with no partiality towards Dr. Sharpey; but the evidences of his talents, of his erudition, and of his accurate acquaintance with general anatomy and physiology were more than to overcome every doubt. In all elections it is undoubtedly the interest of the Professors and the Council to divest themselves of private feeling - to allow no personal friendship - no party connection, and no professional interest to predominate in the choice, but to be guided solely by the good of the establishment. Upon such principles the election of Dr. Sharpey has been conducted". This effectively closed the matter and the Lancet had nothing more to say but, as will be seen, Sharpey remained a target for its vituperation on several other occasions.

## IN POST AT UNIVERSITY COLLEGE

Sharpey's work as a teacher will be considered later; the rest of this chapter concerns his administrative activities and other aspects of his career at University College London.

Although not overtly ambitious, Sharpey did give thought to an eventual move from London to Edinburgh, a city for which he retained a lasting affection. He saw the chair of anatomy as a possible goal and he was encouraged to keep this in mind by Allen Thomson as early as 1836 who remarked that taking the London post "increased his chances of the Edinburgh chair". Thomson raised the question again in 1842 to which Sharpey replied cautiously, saying that it would depend upon the circumstances at Edinburgh and London but by no means rejecting it. In December 1845 Thomson was able to inform his friend that Alexander Munro the third was about to retire and if he was interested in the post he should assemble some prestigious referees. In his reply Sharpey carefully considered the pros and the cons of the two institutions in terms of salary and academic prospects. London had better resources but he saw the need to have additional earnings from also having a clinical post or insurance consultancies. Looking ahead to retirement, he recognised his inability to save money, as became evident when he did retire, and he concluded that Edinburgh was really the place for him. The patronage of the Chair of Anatomy at Edinburgh was in the gift of the City Council and so Sharpey wrote to the Lord Provost on the 4th of February to say that he would accept the post if it were offered to him provided only that he first informed the College Council. On the 6th he received a letter from Quain urging him not to make an irrevocable decision. Quain was anxious to keep Sharpey in London and to do so he volunteered to give up some of his classes so that Sharpey could be given an additional emolument. There must have been much personal persuasion for on the next day Sharpey agreed to withdraw his acceptance and he wrote to the Lord Provost accordingly. He also wrote to tell Syme of his decision, expressing his heart-felt regrets at letting down his friends by his vacillation. He wrote of his attachment to Edinburgh where he would have had "the opportunity of passing the remainder of my days in the scene of my early exertions among attached & old friends and in a most honourable situation". He had previously said that he was comfortable and satisfied in London and now that he was to receive a considerable increase in his salary he decided to stay.

His love affair with Scotland as a place in which to live and work, had not deserted him completely since a few years later he thought about applying for the Chair of Physiology at Edinburgh. The holder, John Hughes Bennett (1812-1875), had succeeded Allan Thomson in 1848 and, like Sharpey, he had enthusiastically promoted microscopy and animal experiments, in the curriculum, but, unlike Sharpey, he harboured clinical ambitions. When Alison retired as Professor of Medicine Bennett expressed his interest in the post and this rekindled Sharpey enthusiasm to return to Edinburgh.

He already knew Bennett but they were not close friends. On one occasion (31st August 1848) he had remarked to Thomson that Bennett "seems to have notions of the Chair which in my opinion are rather suited to his own convenience than the requirements of the University". When the Chair of Medicine came up Bennett was in London; Sharpey "was asked to meet him today at dinner as he particularly wishes to see me – But as I wish particularly not to see him, that is not to involve myself by any talk on the matter, which might afterwards give rise to misunderstanding & require explanations – I have excused myself". Sharpey showed a nice sense of what was appropriate in the circumstances. In the event, as Sharpey predicted, Bennett did not impress the electors and he continued to hold the Chair of Physiology until 1874. Sharpey was sanguine about the business; as he told Thomson:"I made up my mind that if an opportunity of moving to Edinburgh offered itself I would not let it go past – but should there be no opening I am quite contented".

He remained at University College for the rest of his career without, so far as is known, considering any other moves and when he retired he chose to stay in London rather than go to Scotland.

## THE COOPER-SYME AFFAIR

During his years at University College, Sharpey became increasingly involved with College business and through his easy manner and sound judgement he came to have a strong influence in its counsels. This inevitably brought with it skirmishes with colleagues who felt aggrieved over the outcome of issues which affected them. One such quarrel irrupted over the appointment of James Syme as Professor of Clinical Surgery and the resignation soon afterwards of Samuel Cooper (1780-1848), senior Professor of Systematic Surgery.

Cooper had qualified MRCS in 1803 and served in the army from 1813 to 1815, being present at the battle of Waterloo. He was appointed to his chair in 1834, was President of the Royal College of Surgeons in 1845 and elected FRS in 1846. He was the author of two well-received books on surgery and, according to the Lancet obituary, "as a teacher he was greatly esteemed and as a friend and councillor he was greatly loved". During his later years, on the grounds of poor health, he had been assisted by Robert Liston, Professor of Clinical Surgery and Surgeon to the hospital who himself, had to retire through ill health in November 1847 (he died in December). To fill the vacancy immediately, Council agreed to Cooper's suggestion that his son-in-law, Thomas Morton, then on the surgical staff of the hospital, should be appointed on a temporary basis without any commitment with regard to Liston's eventual successor. This latter appointment was discussed by the Senate on the 17th December and it was recommended to the Council that Syme, then Regius Professor of Surgery at Edinburgh, would be a very acceptable colleague. At the Council meeting the next day there was read a letter dated the 4th December from Syme to Sharpey indicating that he would accept the post if it were offered to him. Syme was therefore appointed as from the 8th January 1848 with the same arrangements for helping Cooper and undertaking surgical practice.

On the 6th April Cooper ended his lecture course with the surprising announcement that he was going to resign forthwith. This was not, he said, due to ill health or even to the difficulty he was experiencing now that Liston was gone, but due to "the impossibility of any agreement between me and two of my colleagues (the two who almost rule the medical end of this institution) on certain points affecting the claims of gentlemen brought up at this school, not to be forgotten in the distribution of its patronage". In short, he was dissatisfied with Syme as Liston's replacement and disgruntled because Morton, the insider, had not been appointed instead of Syme, the outsider. The two colleagues referred to were, of course, Quain and Sharpey.

When Syme came to London he found, to his surprise, that Morton was already in position as Cooper's assistant and then, upon Cooper's untimely resignation, Council asked Syme to take on the whole of Cooper's course as well as his own. This would have severely restricted the time available for clinical surgery and Syme felt that the College had not abided by its agreement. A final blow to his confidence in moving to London was essentially trivial but it was the spark which ignited the bombshell of his resignation after only five months in office. The circumstances were as follows:

at a prize giving ceremony on the 7th May the appearance of Quain and Sharpey on the rostrum was greeted with barracking and hissing by the students which the Chairman, Lord Brougham, found difficult to control. Syme wrote later that he had "witnessed a most painful scene in the contumelious treatment of two gentlemen standing to me as colleagues. One of these was a very old friend (Sharpey) for whom I entertained the greatest respect and most sincere regard, who has devoted no ordinary talents ....to the services of a school, in his zeal for which he declined a chair in anatomy, yielding more than double the emolument of that which he now occupies besides being in other respects more advantageous".

Syme returned forthwith to his former chair in Edinburgh; his resignation was a blow for the medical school and a great personal disappointment to Sharpey.

As might be expected, the whole affair became a matter of public debate. The Lancet, ever alert to any opportunity to attack University College or Sharpey, weighed in as the champion of local candidates and open competition and the College Council was put in the embarrassing position of defending the actions of two of its members, Quain and Sharpey.

The Lancet's attack opened with the challenge that Cooper's replacement should have been Morton: "it will not be tolerated that the gentlemen who have been educated at University College should be excluded from the vacant chair". It launched a diatribe against outsiders who had obtained their positions through nepotism and intrigue, referring in particular to Syme's appointment as being due to " a gross piece of Scotch jobbing" by Sharpey and asserting that "At the present time Mr. Quain's position is peculiar and suspicious".

Cooper's resignation address was printed in the Medical Gazette and was followed a week later by a well-argued reply from Sharpey who pointed out that Morton fully understood the provisional nature of his appointment and that Cooper's strictures against "sending for strangers" rather than appointing alumni, did not stand up to examination; out of fifteen appointments made over the previous twelve years, ten had gone to alumni. Cooper had not objected to Liston coming from Edinburgh to take the chair neither had he objected to any other appointments which had been made except for that of Syme. It was clear, he said, that Cooper was motivated solely by his wish to see his son-in-law promoted. The editorial in the same number of the journal sided with the Council saying that the system of "sending for strangers" had worked well for the College; the Council, it maintained, "must select the fittest man, even although he may have the misfortune to be a Scotchman or an Irishman", (written in all seriousness, without the levity of an exclamation mark).

The Medical Gazette rapidly closed all further discussion of the subject but not so the Lancet; this type of affair was meat and drink to it. It reprinted Sharpey's article thus allowing Cooper to return to the fray. He clearly saw a conspiracy aimed at installing Syme, Sharpey's old friend and fellow countryman, into one of the chairs and then manoeuvring Quain into Cooper's chair thus reinforcing their power in the medical faculty. He was particularly wounding in accusing Sharpey of "being a more important visitor in the College office than in the dissecting room". Quain replied in the same issue with a moderate but factual rebuttal of Cooper's charges and an expression of regret at the "personal collision with Mr Cooper, with whom for a series of years I had been on habits of very close intimacy", brought about entirely by Cooper's strident support for his son-in-law Morton. The Lancet, in editorials and anonymous letters, continued to fulminate against Sharpey in terms which displayed what today might be considered as racial prejudice. The office of surgeon went to " a gentleman from north of the Tweed" through the Scottish influence on the Council which made room for "a stranger from Edinburgh". (But the editor, like Cooper, found nothing amiss with the appointment of Syme's predecessor Liston, also a Scot).

The three protagonists, Cooper, Quain and Sharpey, circulated memoranda stating their positions summaries of which were reprinted in the Lancet, to keep the pot boiling.

Sharpey set out to cool the atmosphere: "I have no wish to strive for the last word, still less to keep up animosity but too much fomented already; and although I have felt aggrieved by the way he first took to make known his complaint, and by the imputations which it conveyed, I trust in what I have to say I shall not forget the respectful consideration which is due to his age [he was 68] and his station as an eminent member of the medical profession, as well as to his past services as a Professor of the College". However, he does not let Cooper's seniority be an excuse for personal abuse, such as "meddling professors" and their "dictatorial sway"; "to call names is not the same thing as to prove them to be deserved … and I hope that Mr

Cooper's better judgement and returning good taste will yet bring him to see that it is at best but a sorry resource in a controversy". He concluded that "In his disappointment he raises a storm by throwing up his office, and gives out that it is forced from him by intrigue and annoyance. He doubtless reckons that the positive fact of his resignation in displeasure, and the natural sympathy with an aged professor who retires under any circumstances will be sufficient to establish the case he puts against me; nevertheless, his case (although he may believe it) is not true". Although pacific by nature, Sharpey was able to mount a robust defence when his integrity was questioned.

On the main issue of whether he had used underhand methods to obtain the appointment of Syme, Sharpey denied that there had been any "irregular or censuarable proceedings". It is true that he had made an approach to Syme <u>before</u> Senate had considered Liston's successor and no doubt he and Quain lobbied hard for Syme, but this was done openly and others were entitled to do the same (if they were able to challenge the dominance of their two colleagues). But the acquisition of Syme would have greatly enhanced the prestige of the school; he already held a Regius chair and was at the height of his fame as a surgeon, moreover, unlike Morton, he was unrelated to his sponsors, unless one considers the Scots to be one large conspiratorial family!

It is highly unlikely that a man in Syme's position would have responded to a competitive advertisement; as Sharpey remarked, "considering the position then held by Mr. Syme in the University of Edinburgh, the only way of obtaining his accession to our school was by invitation". He might not have responded to that had he not received prior encouragement from Sharpey who can certainly be seen as acting in the best interests of the school.

Quain's pamphlet was also well-argued and, in view of his reputation, not particularly aggressive; "it is not my purpose to attack Mr. Cooper" he wrote.

The warfare was brought to an end by the College Council which rejected Quain's call for an investigation as being neither necessary nor advantageous. It put down Cooper's charges as being an entire misapprehension, adding that "Mr. Cooper has been strangely imposed upon if he has been informed that any Member or Members of the Council have been influenced, directly or indirectly, by any of the Professors, or any other parties, in the course pursued". It concluded by exonerating Quain and Sharpey "whose talent, character and positions preclude the idea that they could be guilty of caballing for any unworthy purpose".

Cooper's death in December was seen by the Lancet as being hastened by "the insulting and iniquitous proceedings of the College Council which led to his retirement and to which his sensitive and honourable feelings were then subjected. Mr Cooper lived to see the result of these proceedings in the ruined character and diminished means of the institution", this being more the wish of the editor than what actually happened.

Events moved on; Quain, the only winner in this battle, took the chair left vacant by Syme's resignation. Cooper's chair was filled by James Moncrieff Arnott (1794-1885) (not to be confused with Sharpey's friends Neil Arnott or his brother James Arnott mentioned previously), but another Scot, from Fife, suspiciously near to Arbroath and an Edinburgh graduate to-boot; however the hidden hand of Sharpey has not been detected in this appointment.

Sharpey wrote about the Cooper - Syme affair in his letters to Allen Thomson. Firstly, in optimistic vein in December 1847 he says "I truly sympathise with you in your deprivation of Syme's services, but your loss is our gain - He comes to us in our greatest need and the handsome way in which he has behaved in all the necessary arrangements demands our best efforts to second him both in his capacity of clinical teacher and practising surgeon". But writing on the 18th May, when it was known that Syme was to resign, he gives an insider's view of events:

> "You will have heard that Cooper raised a storm against us, which with the malicious comments of the Lancet had the effect of causing some effervescence on the occasion of the public distribution of prizes - Still the tokens of disapprobation directed against myself and Quain on that day (for there were none against Syme who was well received) were mixed with a larger proportion of manifestos of a favourable kind - The hisses came mainly from some of the men on Town formerly students but not now attending and the whole affair would I daresay passed over with little notice had Quain held his tongue.
>
> For my part I cared not much about it because I felt confident that it was merely the safety valve and when the malcontents had cut their hiss the whole affair would soon be forgotten

and things go on smoothly. But as you know Syme has taken
the very inaccountable step of retiring....he gave neither the
Council nor anyone connected with the College any opportunity of explanation for he sent in his resignation and applied to
the Lord Advocate for the Edinburgh chair before he let me
know what he was about. On my endeavouring to reason with
him he generally ends by telling me that the step is taken irrev-
ocably and that it is no use to say more".

He goes on to say that Syme had no complaint against his colleagues
or his students and there was every indication that his appointment would
be a great success. Since it was not to be, Sharpey generously expressed
his happiness that Syme had been able to re-establish himself in Edinburgh
without suffering financially or professionally by "his untoward experiment".

A final word should be said about Morton who was unintentionally
caught up in this affair. He had qualified at University College in 1835 and
immediately became an assistant surgeon there, the first 'home bred' student to do so. He spent some time as a demonstrator in anatomy under
George Viner Ellis (1812-1900), Quain's successor, and after Syme's resignation he was appointed full surgeon. Unfortunately he became an alco-
holic and his behaviour had given offence on more than one occasion. His
father-in-law, Cooper, who died in December 1848, was aware of all this
and had arranged to place any inheritance outside Morton's control.
Apparently Morton was an able surgeon and an excellent teacher but he
had a shy and sensitive nature and in a fit of depression, fuelled by alco-
hol, he took his own life in October 1849. Sharpey recalls this in a letter to
Thomson later that year in which he displays a somewhat censorious attitude to Morton's plight, perhaps a reflection of a more puritanical Scottish
upbringing. "I don't doubt", he says, "that the consciousness of hopeless
slavery to a degrading propensity was at the bottom of his despondency"
and he refers to "Morton's pernicious habit" which was long standing and
not brought on by his failure to obtain Cooper's post.

## LATER YEARS

For the next twenty seven years Sharpey had a busy life, dividing his
time between his teaching duties at University College, untroubled by any
further disputes, and looking after the affairs of the Royal Society after he

was elected Secretary in 1853. He conscientiously pursued his lecturing and examining (he was a University examiner from 1840 to 1863) as generations of students passed through his hands to increase the number on whom he made his mark. His work as a professor and at the Royal Society left him with little time for other interests and although he became Dean of the Medical Faculty and on the Board of Governors of University College Hospital, he had little to do with the profession and he never practised. He was aware of the dissatisfaction felt by many doctors and politicians of the range and varying standards of qualification and of the domination of the profession by the Royal College of Physicians and Oxford and Cambridge Universities. This led to a concerted movement for reform, shared to some extent by the government which, in 1834, set up a Select Committee on Medical Education. Although evidence was collected no recommendations were made since the Committee was dissolved upon a change of government. Not being a topic likely to lead to political advantage or civil disorder, medical reform wound its way slowly through committees, reports, speeches and pamphlets. One such was written by Richard Quain in 1845 entitled *Observations on the education and examination for degrees in medicine, as affected by the new Medical Bill* about which Sharpey wrote:

> "I expected no other reception for Quain's Pamphlet – it may be open to the charge of minor inaccuracies, but the real cause of offence in certain quarters is its truth. Of course it is unpalatable to the old Faculty who resisted pertinaciously for 15 years the opinions of all judicious men without its pale in the matter of preliminary education, and who admitted Midwifery & genl. Pathology into the medical curriculum only on compulsion".

The nature of preliminary education, ie. what subjects should be studied before clinical medicine and surgery, was one of the main issues of the reformers in England. Oxford, Cambridge and the Royal College of Physicians clung to the view that the higher echelons of medicine were the province of cultured gentlemen and therefore an arts degree, combined with residence in a College, were the only essential requirements. The reformers, especially those who had qualified in Scotland, sought a more relevant, scientific basis for clinical work. Sharpey fully concurred with this view and when on the General Medical Council he was closely involved with the establishment of a common pre-clinical course for all medical schools but that was not until the passing of the Medical Reform Act of 1851. He himself was not a campaigner for reform, feeling; as he remarked to Thomson in July 1857, "I am quite out of the medical world so I am far behind as to what is doing or to be attempted by the universities in the medical Bill being in committee. I always feared that the influence of the Corporations would be sufficient to carry their measure".

In Edinburgh, Syme, too, was calling for reform in a letter to the Lord Advocate in 1849. Sharpey agreed with his friends views but, being a realist, he pointed out that: "Could the government deal with existing interests as Napoleon or Nicholas [ie. as dictators, of France and Russia respectively] - the matter would have been settled long ago. Syme forgets that the Lord Advocate is not the Parliament".

He returned to the subject in 1863 when he advocated the establishment of a single licensing board for each of England, Scotland and Ireland, and again in 1872 when he gave his support for the English Conjoint Examining Board set up by the physicians and surgeons the previous year.

It is clear that Sharpey was not actively involved in medical politics; he was quite happy to get on with the tasks in hand, which kept him fully occupied for the rest of his career. His undemonstrative but warm-hearted efficiency was recognised by many of his colleagues and old students when, in 1869, they showed their admiration and respect by collecting a subscription for a suitable memorial to him. The sum of £2170 was raised of which £420 was spent on a portrait by J. P. Knight (which was not a good likeness) and a marble bust by W. H. Thornycroft. In 1871 the balance of this Sharpey Memorial Fund was turned into an endowment for a Sharpey Physiological Scholarship for research under the directorship of the Jodrell Professor. Appropriately, the first scholar to be appointed was his pupil Edward Schafer who himself went on to become one of the leading physiologists of his time. The list of Sharpey scholars during the first fifty years (Note 8) shows how well physiology has been served by this endowment. Almost without exception the scholars have gone on to have distinguished careers as professors or similar in Universities of repute and eight of the twenty one were elected FRS, a record which would have given Sharpey great satisfaction. The scholarship continues into the present day and the holders show the same distinction as their forebears; in their hands the future of physiology is assured.

Another of Sharpey's long-standing friendships should be recorded since it spanned his entire career. Robert Willis (1799-1878) is remembered for his scholarly translation from Latin of the complete works of William Harvey, published by the Sydenham Society in 1847. It is the only edition of all that Harvey wrote and it is still regarded as a standard text despite several later translations. Willis was a fellow Scot, the son of a Leith merchant, and also a medical graduate of Edinburgh, in 1819. Like Sharpey, he spent some time furthering his medical education in Paris and elsewhere. Writing about him in 1842, Professor Karl Marx of Gottingen used the same phrase to describe his travels as Allen Thomson did, either by coincidence or a lapse of memory, forty years later when writing about Sharpey. Marx wrote: "Robert Willis, a physician, who in his youth, *with knapsack on his back*, [my italics] made a walking tour through Germany,....a sensitive, thoughtful, truth-loving man", as could have been said about Sharpey. Willis's career, again like that of Sharpey, brought him to London, initially to practise as a surgeon, but his interests lay elsewhere and in 1827 he was appointed foundation librarian at the newly established Royal College of Surgeons of England. He was energetic in book purchasing and in the organisation of the new library and by 1831 he had compiled its first catalogue, using a modern system of classification. Through his efforts the College came to possess one of the greatest historical medical libraries. Possibly because of ill health, he retired from this post in 1845 and took up general practice in Barnes, south-west of London. He still found time to write a number of medical books among which was *William Harvey: A History of the Circulation of the Blood* (1878). This was his last book and was in importance second only to his translation of Harvey's works; it was dedicated to 'his old friend' William Sharpey. He was buried in St. Mary's churchyard along with his wife.

The friendship may have started when Sharpey was still a student at Edinburgh, two years behind Willis, or later in Paris where they both spent some time. There was also a connection through their mutual friend James Syme who Sharpey first met in Paris. Willis's sister Anne married Syme in 1834 and his other sister, Fanny, has already been mentioned in connection with a social tiff in Edinburgh in which Sharpey played a part. A further small link is that Lister, one of Sharpey's best pupils, married Symes' daughter Agnes, the niece of Willis. Nevertheless, there is not a great deal on record to connect the two men except some remarks by Sharpey in his letters to Allen Thomson. He mentions that on first coming to London he shared a house at 25 Dover Street, Piccadilly, with Willis, where, no doubt, the two academics, professor and librarian, cemented their friendship through their shared love of literature. Soon after arriving in London Sharpey remarked that "I am ruining myself buying books and mean to have a complete set of French and German periodicals immediately connected with my subject". He would have been pleased that Willis undertook a translation from German of Wagner's *Elements of Physiology* (1841) which added to the meagre stock of English text-books on the subject. He commented "Willis has translated Wagner's little volume and I understand that Todd [R. B. Todd of Kings College] having done a little of it will *father* it all, so that it will appear with the Priest's arms on the title page", a nice perception of a colleague's character and a pithy reminder of the distinction between Kings and University College. (Sharpey was wrong since there is no reference to Todd and no coat of arms but the comment illuminates his personality).

Sharpey and Willis had much in common: national origins, university, profession, a knowledge of languages (Willis translated from Latin, French and German), an unworldly lack of ambition and, above all, a dedication to academic values. Although neither was a genius, both were thorough and energetic. They were well matched; it would be interesting to know how their lives interacted. Willis differed from his friend in one major respect: he was married and had a family of six. One of his daughters sketched the only likeness we have of her father, now in the Print Room of The British Museum.

Sharpey retired from his chair in 1874 but he continued to live in London where he had many friends and interests after 38 years residence and where he could continue to enjoy the social amenities of the Athenaeum Club to which he had been elected in 1846. He was joined in 1877 by his closest friend Allen Thomson which must have given him much happiness. He continued to take an interest in University College as shown by a letter he wrote to Schafer in 1877 in which he expressed concern about the need to improve the medical school in the face of competition from the other London schools. "Barts is spending £50,000 on theirs — University College depends much upon the support it receives from old students who could

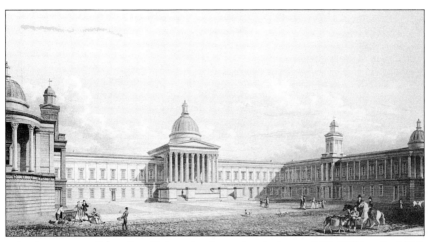

17. The proposed London University from the designs of W. Wilkins in 1825. *Engraving by Thomas Higham. (Courtesy of College Art Collections, University College London)*

18. London University as it was when Sharpey arrived. By C. W. Radcliffe *from C. Johnstone* Public Buildings of Westminster Described *1831.* *(Courtesy of College Art Collections, University College London)*

19. Old Gower Mews, close to London University in 1835. Water colour by G. S. Shepherd. (*Courtesy of College Art Collections, University College London*)

20. Lithograph of William
Sharpey by Alexander
Blaikley 1838.
*(Courtesy of Special Collections, the
Robinson Library, University of
Newcastle-upon-Tyne)*

21. Drawing of William
Sharpey by T. Bridgford.
*(Courtesy of the National Library of
Medicine, Bethesda, USA)*

*22. William Sharpey about 1845. (Courtesy of the Wellcome Library, London)*

*23. Portrait of Richard Quain by T. Bridgford. (Courtesy of the Wellcome Library, London)*

*24. Mezzotint of Richard Dugard Grainger 1827, after T. Wageman. (Courtesy of the Wellcome Library, London)*

*25. Lithograph portrait of Robert Grant by T. Bridgford. (Courtesy of the Wellcome Library, London)*

*26. Samuel Cooper,*
*lithograph after T. Bridgford.*
*(Courtesy of the Wellcome Library,*
*London)*

*27. James Syme, wood*
*engraving after J. Moffat.*
*(Courtesy of the Wellcome Library,*
*London)*

*28. Robert Lee, wood engraving by Smyth, 1851, after Mayall. (Courtesy of the Wellcome Library, London)*

*29. Thomas Snow Beck.*
*(Courtesy of the Wellcome Library, London)*

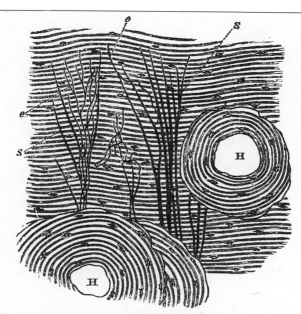

FIG. 68.—Transverse section of decalcified human tibia, from near the surface of the shaft.

H, H, Haversian canals, with their systems of concentric lamellæ; in all the rest of the figure the lamellæ are circumferential; s, ordinary perforating fibres of Sharpey; e, e, elastic perforating fibres. Drawn under a power of about 150 diameters.

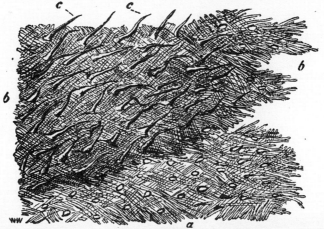

FIG. 69.—Lamellæ torn off from a decalcified human parietal bone at some depth from the surface. (Sharpey.)

a, lamellæ, showing decussating fibres; b, b, thicker part, where several lamellæ are superposed; c, c, perforating fibres; the fibrils which compose them are not shown in the figure. Apertures through which perforating fibres had passed are seen, especially in the lower part, a, a, of the figure. Magnitude as seen under a power of 200, but not drawn to a scale. (From a sketch by Allen Thomson.)

*30. Sharpey's fibres, from Schafer's* Essentials of Histology, *4th edition, 1894.*

send their sons and apprentices — This connection is very important and no means of strengthening it should be neglected".

William Osler (1849-1919) after qualifying in Canada in 1872, went to work with Burdon Sanderson for 17 months and he recalled that "Dr. Sharpey had resigned the previous year but was much about the laboratory and often came to my desk in a friendly way to see the progress of my blood. One evening he asked me to dinner; Kolliker, Allen Thomson and Dohrn were there. When saying goodbye he gave me Davy's Researches with an autograph inscription". This says much for Sharpey's unfailing encouragement of the young and also his generosity; Osler was only 26 and it might have been somewhat daunting for him to be mixing socially with such eminent elderly men. He was at that time writing what is considered to be one of the best accounts of blood platelets (first described by Donné in 1842) a subject in which Sharpey would have taken a keen interest.

Money had never been an important consideration for Sharpey; he lived modestly but without denying himself the pleasures of travel and book buying (no doubt at the medical bookshop of H. K. Lewis which had opened, very conveniently, in Gower street in 1844). He was generous by inclination and he knew that he was not one of nature's savers. The Edinburgh appointment had appealed to him because of the better provision for retirement but it was not important enough for him to make the move. He was not careless about his finances and on one occasion he remarked indignantly to Thomson about "the rascally Tory income tax" when it was reintroduced by Peel in the budget of 1842.

Writing closer to retirement he considered what funds he would have: his professorial salary of £100 should continue, he had £250 a year from investments and he received an allowance of £200 a year from his nephew Major William Colvill while still on active service; in total a very satisfactory income for the times, as he was well aware. His colleagues were, however, unsure about his means and John Storrar (a fellow member of the Senate and the General Medical Council) wrote to Thomson about it saying that "As a member of the College Council I can say that no Professor is more respected by them and that they would one and all be anxious to make the remainder of Sharpey's days easy". As a result his two friends Thomson and Quain exchanged letters on the subject. Thomson wrote "I know too the sacrifice he made in declining the Edinburgh Chair of Anatomy, in the possession of which he would have been wealthy beyond most scientific men". Quain, still a man of influence, mentioned the matter to Robert Lowe, Member of Parliament for the University of London and a former Chancellor of the Exchequer and to Lord Granville, Chancellor of the University, with the result that Gladstone agreed to provide a pension of £150 from public funds in recognition of Sharpey's outstanding services to teaching and administration. On receiving this welcome news from Lord Granville, Sharpey expressed his gratitude to Thomson in a letter of the 24th January 1874 writing: "I suspect that you and Dr. Quain must have first moved on the matter".

The Council of the College recorded its appreciation of Sharpey's services over thirty eight years, awarded him a pension of £100 per annum for life and conferred upon him the titles of Honorary Director of the Museum of Anatomy and Emeritus Professor of Physiology and Anatomy, as recommended by the Senate.

Sharpey remained active during his retirement. He gave advice to the government on scientific education and the use of animals in laboratories (mentioned in more detail later) and he spent time cataloguing the extensive library which he had presented to University College. Some indication of the wealth of his collection is given by Charles Singer in his book *The Evolution of Anatomy*: "Sharpey's own copies of Vesalius, of Fallopius, of Columbus, of Casserius, of Spigelius, of Harvey, and other Paduan anatomists lie open before me as I write. They bear upon them the evidence of Sharpey's vivid consciousness of the antiquity and dignity of the line from which he was descended". The catalogues which he prepared have now come to light at University College. There is a shelf catalogue and a three volume alphabetical catalogue, in his own hand, which shows the breadth of his library. In addition to the works referred to by Singer, there were editions of Graaf, Darwin, Gilbertus, Newton, John Dalton and the Hunters. Among modern authors were Bernard, Poiseuille and Matteuci whose latest books were included. It was an impressive benefaction but the books have now been distributed within the College library and there is no identifiable Sharpey collection.

He was also instrumental in obtaining for the College the library of his colleague Robert Grant who, Sharpey discovered, had not made a Will despite being aged 81 and quite seriously ill. Sharpey and Ringer visited him early in August 1874 and urged upon him the desirability of having a solicitor in to take down his last wishes. Grant would not do so but he did agree to sign a declaration of intent, as Sharpey informed the College Council in a letter of the 22nd August. Grant died on the 23rd August; he had no close relations and by his declaration, which was, in effect, a legal Will, he left to the College, for which he had a deep affection, nearly all his property including his collection of comparative anatomy and zoology specimens and his extensive library. This, like Sharpey's library, has been widely dispersed and in the absence of a catalogue there is no means of knowing what it originally contained. Since Grant was a devoted scholar it is very likely that his library would have included all the literary treasures of his subject. His specimens formed the basis of what is now the Grant Museum of Zoology and Comparative Anatomy which thus perpetuates his name in the College.

From about 1870 Sharpey began to have problems with his sight through the growth of cataracts; he had operations to remove them but they were not entirely successful.

In London he had lived for a long time at Gloucester Crescent, Regents Park, with his then widowed sister Elizabeth Colvill and her children Mary and William. Elizabeth died in 1855, William took up medicine and Mary remained to look after her uncle. In 1867 they moved to Hampstead where they stayed until 1872 when Mary became ill and returned to Arbroath where she died in 1878. For a time he was looked after by his half-sister Jacobina until he moved to lodgings in Torrington Square, conveniently close to University College, but he spent part of the winter in Hastings. On returning from there at the end of March 1880 he developed bronchitis and died on the 11th April after a short illness; he had been attended professionally by Sydney Ringer and his close friends Allen Thomson, John Marshall and Burdon Sanderson were present at his death.

His body was taken to University College from where, on the 15th April, it was borne in a long and impressive cortege to Euston Station for its journey to Arbroath. Schafer wrote that "the procession was so imposing that people passing along the street must have wondered what manner of man it was who was able when dead to draw so many after him".

In his Will, for which John Marshall and Allen Thomson's son John acted as trustees, he made a few small bequests, one to his godson William Sharpey Seaton, son of Edward Cator Seaton (1815-1880), a friend from his Edinburgh days and Medical Officer at the Board of Health; the residue passed to his nephew William Colvill The bulk of his modest estate, £800 in railway stock, was "bequeathed to University College to be held in trust for the purpose of applying the interest thereof in augmentation of the income of the Sharpey Physiological Scholarship"; a lasting memorial to his devotion to learning and his unswerving loyalty to University College London.

*31. The playground of University of London School adjoining the new University building in 1833. Lithograph by George Scharf. (Courtesy of College Art Collections, University College London)*

CHAPTER 3

# *The Royal Society and Elsewhere*

THE ROYAL SOCIETY OF London for the Promotion of Natural Knowledge, its full title, was founded in 1660 and given a Royal Charter by Charles II in1662 but, unlike the academies of science in many countries, it was never a government institution. It received limited financial support, the provision of meeting rooms and the expenses of activities in the national interest such as geographical explorations or navigational observations and it was often called upon for advice on scientific matters. Despite the galaxy of scientific stars in the Society they were never a majority of its members; peers, bishops, men-of-letters (Byron and Tennyson, for example), politicians and often those without any particular status were admitted so that by the beginning of the 19th century the Society could be regarded as one of the better gentlemen's clubs in London. Proposals for membership could be made at any time of the year simply by posting in the meeting room a certificate signed by only three Fellows; election followed almost automatically after it had been displayed for ten weeks. There was no upper limit to the number of members; by 1800 it had reached 685; in comparison the French Academie des Sciences had only 75 members. The Royal Society was governed by a Council, appointed by the President, and administered by two Secretaries, serving either biological or physical aspects of business, and a Treasurer.

From early in the 19th century there grew a growing dissatisfaction by many of the scientific Fellows with what they saw as the declining influence of the Royal Society in an increasingly scientific world. Only a few of the Fellows contributed to the Philosophical Transactions, its hitherto prestigious house journal. Matters came to a head in 1830 with the publication of a number of outspoken attacks on the way the Society was run.

Not surprisingly, the *Lancet* involved itself, maintaining an implacable, and often intemperate, opposition for nearly twenty years. Charles Babbage FRS (1791-1871), mathematician and pioneer of the calculating machine, who had been a secretary of the Society from 1820-1824 and therefore in a good position to understand its affairs, spoke out for reform in his *Reflections on the Decline of Science in England and its Causes* (1830). He exposed the malpractice of some of the officers, called for a reduction in the number of Fellows and for greater openness in the conduct of its business. He pointed out that, by failing to provide the best environment for serious scientific discussion, the Royal Society was driving groups of scientists to found there own specialist societies such as the Linnean, the Geological and the Astronomical, a move which would weaken the Royal. He made a comment about the medical profession which, in view of the controversy which was soon to embroil Sharpey, was particularly apposite:

> "The honour of belonging to the Royal Society is much sought after by medical men, as contributing to the success of their professional efforts, and two consequences result from it. In the first place, the pages of the Transactions of the Royal Society occasionally contain medical papers of very moderate merit; and, in the second, the preponderance of the medical interest introduces into the Society some of the jealousies of that profession".

Criticism also came from the astronomer James South FRS (1785-1867) in a pamphlet *Charges Against the President and Council of the Royal Society* and from a book by the physician Augustus Granville FRS (1783-1872), in *Science without a Head or the Royal Society Dissected* first published anonymously in 1830 but re-issued in his own name in 1836 with the more accommodating title *The Royal Society in the Nineteenth Century*. As a result of these pressures a number of new measures were introduced, for example, the anonymity of referees of papers and the formation of committees of specialists, for example the Physiological Committee, of which more will be heard. It was not until 1847 that a new charter was obtained which restricted entry of new Fellows to fifteen a year and gave all Fellows a voice in the Council and the right to view the accounts. This event was commemorated by the formation of a new dining club, The Philosophical Club, limited to 47 members to mark the year of the charter and with the aim of promoting the work of the professional scientist. Sharpey contributed to its proceedings from time to time speaking on, for example, the chromatophores in fish skin or of his experiences in the use of saline transfusions in asphyxia. There was also in existence a Royal Society Dining Club, a purely social organisation, founded in 1743, to which Sharpey was elected in 1853; its historian records that "he was considered to be one of the most genial and attractive members the club ever possessed".

Sharpey was a Fellow well before the last reforms were carried out. He had been elected on the 9th May 1839 with the support of fourteen men (Note 3) well-known in the academic world including his compatriots Robert Grant of University College and Neil Arnott. The citation in common use for applications read: "Mr. A. a gentleman devoted to scientific pursuits, being desirous of becoming a Fellow of the Royal Society, we, the undersigned, do from our personal knowledge, recommend him as worthy of that honour, and likely to become a useful member". There was no attempt to find out what scientific advances had been made by the applicant.

Sharpey was an active member of the Society. He must have attended many of its meetings and he certainly pulled his weight as a referee of papers submitted for publication; there exists nearly sixty reports by him, mainly on biological subjects but occasionally one on a topic outside his normal field. A significant example was his report in 1850 on the paper by Augustus Waller (1816-1870) on the changes in nerve following section in which he described what was later to become universally known as Wallerian degeneration. Sharpey's comments were quite blunt: the introduction was somewhat superficial and "did not contain anything beyond what is to be found in elementary treatises". He pointed out some relevant references not known to the author, an indication of his wide reading of foreign papers, and concluded that "although Dr Waller has been anticipated to a much greater extent than he must have been aware of....his researches still possess a considerable share of novelty and importance". He recommended publication once the defects had been corrected and so a work of considerable importance became known through the Royal Society.

This was the first of several connections with Waller. He provided laboratory facilities for Waller to use when over from Paris and Waller was a candidate for a lectureship in practical physiology at University College before withdrawing through ill health. A partnership between Sharpey and Waller would have been greatly to the benefit of British physiology. Several years later Sharpey was able to point out Waller's priority over Cohnheim for the discovery of diapedesis, the ability of white corpuscles to move across the capillary membrane into the extra-vascular space. This he did in a letter to the Lancet in 1868 following the publication of Cohnheims work which had made no reference to Waller. Later, as Secretary of the Royal Society, and perhaps at his instigation, he wrote to Waller in April 1870 inviting him to give the Croonian lecture that year.

THE LEE AFFAIR

This bad tempered quarrel between a number of Fellows takes up more space than is justified by its scientific importance but it shows how Sharpey responded to a difficult situation made worse by clashes of personality. It also gives some idea of the fervour and vituperation generated in these conflicts of the last century in contrast to the more moderate standards of academic warfare which exist today.

Soon after his appointment to the Council in 1844 Sharpey became embroiled in a dispute which called into question his integrity as a scientist and, of greater consequence, the integrity of the Council and thereby that of the whole Society. The dispute arose over the award of a Royal Medal for Physiology which, it was claimed, was made both unfairly, because better work had been passed over, and illegally because the correct constitutional procedures had not been observed. The protagonist was Robert Lee MD.

Lee was born in 1793 in the parish of Melrose in the Scottish borders, the second son of a prosperous farmer. He was educated at Galashiels and then at Edinburgh University, initially with a view to entering the church but he switched to medicine and graduated MD in 1814 also becoming MRCS Edinburgh. He held a post at the Edinburgh Royal Infirmary until 1817 when he went to London to take up practice. In 1820 and 1821 he spent some time studying in Paris before establishing himself in London as a physician-accoucheur with the added qualification of LRCP. He had to abandon this owing to ill health and he then spent some time as physician to the family of a Russian count in the Crimea. He returned to London at the end of 1826 and devoted himself to the science and practice of mid-wifery. He lectured at the Webb Street school of medicine and held appoint-ments at a number of smaller hospitals. Through his friendship with Sir Gilbert Blane MD FRS he was introduced to the Royal Society and elected a Fellow in March 1830. In 1834, through the influence of Lord Melbourne, whose family he had attended in London, he accepted the Regius Chair of Midwifery in Glasgow but he resigned within a few weeks, even before giving the public inaugural lecture he had prepared, and returned to London. The next year he was appointed Professor of Midwifery at St. Georges Hospital where he remained for the rest of his working life. He became FRCP in 1841 and gave the prestigious Lumleian, Croonian and Harveian lectures (the latter he attempted to give in the traditional Latin but he had to revert to English when only part way through and thereafter Latin was abandoned).

Lee was a dedicated and industrious investigator who published many papers and a book on clinical midwifery and gynaecology. He was also interested in more academic aspects of his subject in particular the nervous anatomy of the uterus, which will be considered shortly. He produced an excellent description of the nerves and ganglia of the heart in 1849, thereby confirming the earlier work of Remak, and he described the sequence of ovarian changes during the menstrual cycle following on from Cruikshank's observation of the ruptured Graafian follicle in 1797. He was certainly a man of ability with the typical Scottish passion for learning and teaching. Unfortunately he was also cantankerous,intolerant and resentful of criti-cism. He could spoil his case by his abuse. In another dispute the *Lancet*, not short on invective itself at times, remarked that "Dr. Lee would do well to moderate his tone and his language". Lee retired from practice in1875 and died at Surbiton in 1877. His death went unnoticed by the Royal Society and, strangely, by the *Lancet*, but his entry in *Munk's Roll of the Royal College of Physicians* gave a fair account of a career marked by many achievements but marred by his difficult personality.

Lee's dispute with Sharpey and the Royal Society is set out in detail in a *Memoir* he published in 1849, when the dust had settled. It provides the text of letters exchanged between the parties involved and is less one-sided than one might have expected from someone so stubborn and opinionated. The *Lancet* gave extensive coverage between 1845 and 1848 publishing letters and keeping up a flow of scathing editorial comments to feed the flames of the conflagration. It must have made exciting reading for the uncommitted *Lancet* subscriber. If Sharpey's long walk down the groves of academe appears to be a smooth journey, accompanied only by the accolades of his colleagues and pupils, the Lee affair will show some of the stony passages that from time to time threatened the equanimity of that worthy traveller.

The origin of the dispute can be traced back to the late 1830's when Lee was making dissections of the nerves of the gravid uterus; he became utterly convinced of the truth of his observations and he did not take kindly to any criticism. But, as we shall see, he developed a paranoia which went far beyond a robust defence of his work.

At some stage he showed his material to Sharpey who must have expressed some doubts about whether all the observed tissues really were nerves, as Lee claimed. Some portions were given to Sharpey to section and examine microscopically to resolve his doubts. On writing to ask about the results Lee received the following reply dated December 1840:

> "Dear Sir, At this distance in time I cannot venture to put down the result of the microscopic examination of the textures you gave me; but my impression is, that it was not decisive, for one of the most important specimens to be compared was unluckily left behind at your house in transferring the things from one bottle to another, and, if I rightly recollect, what I examined with the microscope were supposed nerves in the broad ligament, about which you were less confidant, and which seemed to me to be slender bundles of tendinous fibres. I may add, that the reasons for which I ventured respectfully to differ from you when you asked my opinion, were quite independent of microscopic evidence; nor do I rely on the test of acetic acid, which appears to me inapplicable to small branches of the sympathetic nerves."

This was a somewhat evasive reply. Mislaying an important specimen; having only "an impression" and being uncertain about his recollections; involving Lee in his doubts; placing more reliance on visual examination rather than on the microscope; these features seem at odds in someone who was normally precise and a champion of microscopy in anatomical teaching. Lee was not convinced and he asked Sharpey to examine the dissections again which he did in September 1841 but with the same result: doubts but not certainty. Sharpey made no further examinations despite being asked to do so; by this time (1842) Lee had published two papers in the Philosophical Transactions on his findings. The two men clashed at a Royal Society soiree in 1842 when Sharpey remarked to Lee, "in a peculiarly abrupt manner", that "in three years time I would give up, or be compelled to forego, all that I had published respecting the nerves of the uterus". This might not have upset a less sensitive person but Lee took it as evidence of a conspiracy to undermine his reputation and all his future actions were coloured by this view. He repeated this remark when writing to the Lancet in November and Sharpey, in his reply, showed that he too could respond in kind. Referring to the charge that some the confirmatory dissections (by Beck, see below) "were made at my instigation for an unworthy end" he goes on "what reliance can be placed in Dr Lee's averments? His imagination seems to warm with his animosity". Recalling the meeting with Lee, he remarks, somewhat tartly that " although, as he correctly says, I had never been honoured with his intimate acquaintance, Dr. Lee yet honoured me, as he was in the habit of honouring many others, by entering into conversation on the never failing subject of his nerves". Sharpey goes on to say that he merely remarked, by way of a joke, that "in two years time he would not believe in his nerves himself".

Another actor had by now taken the stage. Thomas Snow Beck (1814-1877) had been a pupil of Sharpey at University College on the way to qualifying MRCS in 1839. He studied in Paris for two years before returning to University College Hospital as surgeon for which position Lee had supplied a laudatory testimonial. Beck had known Lee from his student days and had visited him at home frequently; he had become conversant with the work on nerves, seen the dissections, and had no doubts that Lee was correct. The subject took his interest and in February 1843 he had the opportunity to dissect for himself a post-mortem gravid uterus, with Lee's knowledge and approval. Lee maintained that he acquired it and offered it to Beck; the latter disagreed about this. Whatever the case, Lee went to see how the work was proceeding and, according to him, found the specimen in "a half putrid state" mangled by clumsy dissection. Moreover, he learnt that Beck had taken the taken the specimen to Sharpey's department and had been given financial aid for his work in the form of a supply of spirit and glass-ware. Lee must by then have known that Beck was supporting Sharpey in his belief that many of the tissues were not nerves. Here, then, was more evidence of a conspiracy. Beck was an accomplice who had been put up to make the dissections which would prove Lee to have been wrong. Beck, once a welcome visitor, was now a traitor. No wonder that Sharpey had been so confident that Lee's work would be refuted within two or three years!

At this stage it would, perhaps, be helpful to summarise the differences between Lee and Beck over the uterine nerves *per se* before the issue becomes clouded by personalities and procedures. These nerves arise from the lumbar sympathetic outflow and join to form the paired hypogastric plexuses from which non-medullated fibres pass to the uterus and adjoining viscera. There is also a supply from the sacral (parasympathetic) spinal cord which unites to form the pelvic nerve distributing branches to the viscera from the pelvic plexus which overlaps with the hypogastric plexus.

There was some disagreement over the innervation of the reproductive tract, Beck maintaining that the pelvic nerve supplied mainly the vagina, bladder and rectum, not the uterus. The main conflict was over the claim by Lee that the nerves of the uterus increased massively in size during pregnancy. He described what he believed was a well-developed network of long and broad plexuses and ganglia over the body of the gravid uterus which amounted to a special nervous system for this organ with the implication that it was important during pregnancy. Beck, on the other hand, found the uterus to have a relatively poor nerve supply, which did not enlarge during pregnancy and which, therefore, might be assumed to play little or no part in the process. Lee had postulated a functional role of nerves in conception, gestation and parturition whereas Beck, and many others after him, found that gestation and delivery could take place in the absence of any nervous connections. He maintained that what Lee had described as nerve was really "a layer of organic muscular fibres in the form of broad fasciculated bands".

At a meeting of the Royal Society in June 1845 Lee read a Supplement to his published papers on nerves. He did this because, as he says, he became "aware that the scheme which had been devised for the nervous tissue of the uterus and, as was expected, of my character as a scientific observer, was now fully ripe for execution".

At the same meeting a paper "On the Nerves of the Uterus" by Beck was read by title, only, which must have been seen by Lee as base ingratitude for the hospitality formerly extended to him.

On the 27th October 1845 the Physiology Committee of the Royal Society met, with Sir William Lawrence (1783-1867) in the chair. (A point to note is that he had been a referee for Lee's two published papers on nerves). Amongst other routine business, Beck's paper was considered and it was agreed that it should be published in view of the favourable report by Sharpey and Todd. Lawrence closed the meeting and departed quickly to keep another appointment; the rest of the Committee had not dispersed when it was realised by Roget (1779-1869), the Secretary, that they should have discussed the award of the Royal Medal; later he said that he had confused physiology with physics for which no medal was to be awarded that year. The meeting was reconvened with Todd in the chair and sixteen papers were considered for the award; one of them was Lee's Supplement. It was decided to recommend Beck for the medal; Sharpey was present and in agreement. Three days later the Council met, formally awarded the medal to Beck and agreed that his paper should be published. Sharpey was later to state that he had not written any of the report on Beck's paper because of his association with the author; he also had not taken part in the Council's award of the medal owing to his late arrival at the meeting.

A week later there was another meeting of the Physiology Committee at which the proceedings of <u>both</u> parts of the previous meeting were confirmed by Lawrence. It was also decided, despite a favourable report by Sir Benjamin Brodie (1783-1862), not to publish Lee's Supplement. Lee soon heard of this and complained to the President, the Marquis of Northampton, that the award of the Royal Medal was not made according to the regulations and therefore should be rescinded. His point was that the re-convened meeting, with Lawrence absent, was not constitutional.

The President replied that the Council alone made the award, independently of the Committee, and having done so the decision must stand. After a further exchange of letters the President asked the Physiological Committee to reconsider the matter and it was then discovered that there was no report by Todd and Sharpey and that, contrary to all the rules, Beck's manuscript had been returned to him for further additions instead of being retained by the Society. An attempt to minimise the damage was made at a later meeting of the Committee when a referees report by Todd and Sharpey was delivered but remained confidential. One member, the botanist J. E. Gray (1800-1875), moved to annul the award but he was outvoted. When the Council met again it received the above report and also letters of complaint over the award from Lee and from Wharton Jones (1808-1891) but nevertheless it confirmed that the award should stand. Possibly as a conciliatory gesture and to end the dispute the Council agreed to publish Lee's Supplement.

In this it failed. In fact it marked a new stage in the battle which continued for over a year longer and which was made public through the intervention of the *Lancet* which used its considerable powers of invective against Sharpey personally (apparently never forgiving him for obtaining the Chair at University College) and against the officers and organisation of the Royal Society, a long standing feud.

Following the Anniversary meeting in April 1846, at which the President referred to the Royal Medal, Lee wrote at length to the Council stating his legal objections and declaring, piously, that "The moral character of the Royal Society requires that a rigorous examination be made of the whole of these extraordinary transactions", a demand that was brusquely turned down by the President the next day, although he did offer to tighten up procedures for awards in future.

By now the *Lancet* was fully engaged and its campaign against Sharpey and the Royal Society was vitriolic and near libellous. It opened on the 28th March 1846 with a letter from George Redford, a London surgeon, who complained that Lee's careful work on many specimens over several years had been unjustly overturned by Beck on the basis of a single specimen. The *Lancet* alleged malpractice in previous awards and declared that Sharpey's involvement with Beck amounted to nothing more than "passing a medal from his right hand to his left". There were also innuendoes about Roget's perquisites as Secretary and the rewarding of favourites; he had long been a target for the *Lancet*.

The next issue contained a long diatribe from Lee to which Sharpey felt he had to reply. He set out his own, accidental, connection with Beck emphasising that he did not instigate the dissections and pointing out his lack of involvement with writing a report or voting for him at Council, much as he agreed with its decision. There was also a letter from Beck who quietly took responsibility himself for the dissections which were made at his house, not at University College, and with the approval of Lee.

The publication of Lee's Supplement and Beck's paper in the same issue of the *Philosophical Transactions* was the stimulus for another outburst of hyperbolic indignation by the *Lancet* but it did go on to make a more rational assessment of the two papers before siding with Lee. The introduction of additional material by Beck after his paper had been accepted, "under pretence of explanation of the plates", was denounced; it was possibly a fair point if made without the insinuation of dishonesty.

Beck wrote again with a detailed comparison between his findings and Lee's and concluding "In February 1843, with a view of confirming the researches of Dr Lee, I commenced the dissection of the nerves of the gravid uterus, but I found so many points at variance with his published statements". These seem to be the honest doubts of an unbiased observer made without any malice towards Lee.

However, Lee was still fighting. He repeated his allegations of conspiracy by Sharpey and Beck and produced nineteen testimonials from well-known medical men who had inspected his dissections between 1840 and 1846. The list included Marshall Hall (1790-1857), William Lawrence, Benjamin Brodie, and that formidable anatomist Robert Knox. This evoked a prompt response from both Beck and Sharpey in the next issue. They denied the existence of the large ganglia, the "brain of the uterus", so important to Lee, but the personal rancour continued and Sharpey showed that he too could use his pen to wound, for example: "Dr Lee either overlooked this fact (an early reference) which should teach him to be indulgent as regards the mistakes of others, or he must be conscious of discreditably concealing it … a practice in which, if some accounts are to be believed, he is not altogether a novice".

Still Lee persisted, perhaps goaded by similar remarks as the above, and the following week he made four charges against Sharpey: 1. Making a judgement without proper knowledge of the subject; 2. Holding an illegal meeting of the Physiology Committee; 3. Making a statement to the Committee unsupported by facts; 4. being party to a violation of the regulations established by Queen Victoria for the award of the highest scientific honours of the Royal Society. Sharpey did not reply to these charges specifically, on the grounds that it was incumbent upon councillors not to be drawn into making personal statements about matters that concerned the Council as a whole. But this did not prevent him from making some outspoken remarks about his adversary: "Dr Lee, in his morbid eagerness to obtain assent to his scientific pretensions, forgets himself so far as to resort to the practice of teasing, reviling and defaming those who differ from him"; "in place of retracting his calumnies and expressing regret for having permitted himself to utter them – to which course a right minded man would feel himself irresistibly prompted – Dr Lee has preferred an ignominious silence".

The *Lancet* persisted in its attack with a vitriolic editorial on the 19th December. Beck was castigated for his outspoken remarks about Lee made, it was insinuated, in order to gain the favour of Sharpey and forward his career. Sharpey's actions were dictated solely by " a *mal animus* entertained against Lee and his labours from the first" and from there the editor launched into his real enemy The Royal Society. "Their secret proceedings make it little better than the Inquisition; they constitute in fact a kind of scientific Star Chamber....we know that papers have been placed in the hands of members of the Council calculated to influence their votes on the merits of a particular paper, and on the question of its publication or banishment, thus mingling spite and pique with what should be a most sacred duty". This was followed by a clarion call for the Society to be "conducted like our open courts of justice; full, free, open and stern justice is all we require".

Lee's last letter in the *Lancet* (19th December 1846) was something of an anti-climax; he merely repeated his four charges against Sharpey, personally, rather than against the Council, and, with some bravado, offered "to forgive him for his personal vituperation" if he would give a satisfactory answer to the charges. Even the *Lancet* was "a little dismayed at the curtness of his concluding note".

This should have ended the dispute but it was re-opened in the following year when Lee, through a resolution by ten of his supporters, forced upon the Royal Society a Special General Meeting on the 11th February 1847 at which he re-told the events at the Physiological Committee in October 1845 and Lawrence made the admission that he had dissolved, not adjourned, the meeting when he left. The President then revealed that the printed minutes of the subsequent meeting of the Council contained no reference to the award of the medal because the resolution had been rescinded and the minute erased. "This illegal and clandestine transaction....thereby rendered the award null and void" thundered Lee. Moreover, the Council itself, at that time, was unconstitutional since membership lapsed automatically at the end of November and there had been no formal re-appointment. Therefore the award of the Royal Medal, for the second time, in December, was illegal and invalid. Despite this admission, the outcome of the meeting was that Lee's motion to rescind the award of the medal was negatived by an amendment by Gray that no further steps should be taken; this was approved by a majority vote. Lee did not accept this and there followed a further exchange of letters with the President who maintained, with great courtesy, that he was unable to alter the decision of a previous Council. It was left to Christie (1784-1865), as Secretary, to inform Lee on the 20th January 1848 of "the unanimous resolution of he President and the Council that it was inexpedient to re-open the question".

This should have ended the matter but Lee continued to resent the way he had been treated and on the occasion of his Lumleian lecture to the Royal College of Physicians in 1855, the *Lancet* saw fit to re-open the affair; this time Sharpey did not re-enter the fray.

A word should be said about Wharton Jones FRS, one of Lee's supporters, who made a late intervention in the affair by the publication in the *Lancet* (13th March 1847) of comments he made at a meeting of the Royal Society in 1846 and of a subsequent letter to the President and Council. This consisted of a detailed criticism of Beck's work together with a jibe at Sharpey for not responding in person at the meeting. This was not Jones's field of interest but he had a long-standing antipathy towards Sharpey which must have played a part in his decision to intervene.

Thomas Wharton Jones,who qualified in medicine in Edinburgh in 1827, was then a lecturer in physiology at Charing Cross Hospital where he taught, and was respected by, T. H. Huxley. He was elected FRS in 1840 in recognition of his work on the micro-circulation and the functions of the ovary, which have stood the test of time. He also became an expert on the eye and in 1851 was appointed Professor of Ophthalmology at University College Hospital thus becoming a colleague of Sharpey. According to his biographer, there was a coolness between the two which stemmed back to their Edinburgh days. Jones had been an assistant to Robert Knox at the time of the Burke and Hare case and had in fact paid Hare for the first body which was delivered. When Knox was hounded out of the city, ending up in London, Jones believed that Sharpey had behaved badly in not support-ing his former colleague but sided with convention (and prudence?). Jones's obituary in the British Medical Journal refers, more forcibly, to "his long and bitter feud with Sharpey which began when they were young men in Edinburgh". There is no doubt that Jones was a difficult character, quick to take offence and caustic about work which did not agree with his own. Sharpey refers to him occasionally in his correspondence with Allen Thomson, on one occasion calling him "that cankered little cat", but he could be generous to him as well. Writing in March 1846 about the chair of anatomy in Glasgow he says: "If you (Thomson) did not get it I should like next to see Wharton Jones in it—he has worked hard and although he has been unwise in getting into scientific controversies which excite people against him, he is a man of great merit".

Jones's letter to the Lancet was answered briefly by Sharpey and Todd, denying any lack of good faith or scientific accuracy over the papers but declining to take any further part in an agitation "set on foot by disappointed expectants and on which the Royal Society had already pronounced".

So ended the Lee affair; who had won? No party came out of it entirely unscathed and Babbage's remarks about the medical profession were thor-oughly vindicated.

Lee showed himself to be rude, intolerant, and consumed with a pas-sion to prove that he was scientifically correct and had been unjustly deprived of the Royal Medal. His treatment by the Society was less than generous if not technically unjust. Beck maintained some dignity and his dissections were not found to be erroneous but, despite becoming an FRS in 1855, he faded from view as a scientist. Sharpey strongly defended his integrity although with a certain lack of frankness in his early dealings and he displayed his personal animosity towards Lee; writing to Allen Thomson in 1845 he declared "Beck has sent his paper to the Royal Society....Lee will be quite floored at least in every body's estimation but his own". The *Lancet* could claim victory in the further reforms forced upon the Society in 1847 and in the resignation of the President and Roget, its *bete noir*, as Secretary. Sharpey was not dislodged but went on to become Biological Secretary. He was still able to evoke a barbed comment from Wakley, its editor, who, on a later occasion, referred to him sarcastically as "Sharpey, the profound physiologist and the autocrat of the elections of the Royal Society". The Society itself must have found the affair embarrassing and distasteful at a time when it was embarking upon reforms which would make it, once again, the most prestigious of the world's scientific academies. The procedural lapses (awarding a medal prior to publication and the illegal adjourned meeting) were, perhaps, symptoms of a casual, amateur attitude typical of a gentlemen's club rather than a deliberate conspiracy designed to harm an awkward and unloved Fellow. But there was a lack of transparency in its relations to Lee even though his findings on the uterine nerves were not upheld.

Lee did not mellow with age and he continued to harbour a grudge against the Royal Society. In 1862, in his 69th year, he read a paper on "Dissections of the ganglia and nerves of the oesophagus, stomach and lungs" in which he concluded somewhat tartly "I have made elaborate dissections…but of these I shall give no account to the Royal Society unless expressly requested by the Council to do so, and assured that my communication shall receive that treatment which I consider the importance of the subject to demand".

By way of a postscript, Sharpey was in touch with Lee in 1863 about a paper submitted to the Society by Lee's elder son Robert James Lee (1841-1924). He had followed in his father's footsteps as a medically qualified investigator via Cambridge, St. George's and St. Thomas's Hospital, London and further study in Paris, becoming MD in 1869 and FRCP in 1873 but, unlike his father, he did not become an FRS. He published papers on a number of clinical and biological topics and he was also an accomplished athlete in his younger days. His paper was written while he was still at Cambridge and was communicated to the Royal Society by his father. It was a description of the nerves of an anencephalic foetus and Sharpey considered that it merited publication after some clarification. He accordingly sent it back to Robert Lee senior, as the sponsor, with an accompanying letter which was the model of secretarial formality:

Royal Society 18th December 1863
My dear Sir
    I herewith send you a proof of Dr Lee's paper which has not yet been read to the Society. On first reading through the MS it

appeared to be a communication suited to be printed *in extenso* in the Proceedings, and I accordingly sent it to the printers to be put up in slip, since then I have read it again with more care, and as the fact communicated in it - of respiration being performed and life for a time maintained in a foetus destitute of a medulla oblongata - is so important, it appears to me to be very desirable that the author should be offered an opportunity of adding a more particular description of the condition of the truncated upper end of the spinal cord and of the connection of the *Par Vagus* with that part; and also of illustrating his description with a drawing of the upper end of the cord and attachment of the nerve, if it can still be obtained.

These additions, I feel sure you will agree with me in thinking, would greatly enhance the value of the Paper; and on consulting one or two members of the Council I found they were of the same opinion. It was therefore thought better to defer the reading of the paper until you had been communicated with in order that Mr Lee might make these additions if he thought fit.

I would also ask Mr Lee kindly to say whether he means the right recurrent nerve really turns round the <u>innominate</u> artery, - for, as he states that the arrangement of the trunk and branches of the right vagus did not deviate from the normal condition, it has occurred to me that the word "innominate" may have been through an oversight written instead of "subclavian".

Sincerely hoping that you will approve of the course taken - I remain

My dear Sir

Yours very truly

W. Sharpey

Dr Robert Lee FRS etc.

The paper was subsequently read to the Society and published in its Proceedings in 1864 but, in the words of the *Lancet* obituary, Lee "never fulfilled his early promise either as general physician or specialist, largely because he was distinctly difficult", a chip off the old block, one might say.

As told, the Lee affair ranks as a burning issue but, except for the one trivial remark, Sharpey did not allude to it in his correspondence with his close friend Thomson; his equanimity enabled him to deal with it as it arose and then to return to more congenial matters.

In 1853 Sharpey was appointed Biological Secretary to succeed Thomas Bell MRCS (1792-1880) who had just been appointed President of the Linean Society and could not undertake both offices. The Physical Secretary at the time was S. H. Christie who retired in 1854, thus making Sharpey the senior secretary, and was succeeded by G. C. Stokes (1819-1903) the physicist, who served for thirty years. Sharpey himself served for nineteen years before being succeeded by his friend T. H. Huxley in 1872 who was succeeded by Sharpey's favourite pupil Michael Foster who served for twenty two years. The Sharpey connection, if it may be so regarded, thus lasted for fifty years.

Much of the recorded business of Sharpey as Secretary was of a routine and unexceptional nature. There is much correspondence about papers for publication, the expenditure of the Government grant and the compilation of a new Catalogue of Scientific Papers, which benefited from his extensive knowledge of the literature. To Thomson he wrote in 1854, not long after his appointment, "I have got along very pleasantly in my post". A hint of what this means is given later when writing about the Anniversary meeting in 1855: at the dinner which followed "we had at table a stupendous pike, three and a half foot long, sent by Mr. Whitbread (1796-1879) the brewer, who is a fair amateur astronomer". It was his duty as Secretary to present reports from time to time and these often included unsigned obituary notices. A most illuminating one was that of Johannes Muller (1801-1858), the eminent physiologist; Sharpey knew him well and it is most likely that he was the author.

In 1857 Sharpey was involved in the move of the Society's rooms from Somerset House to the more spacious Burlington House which it shared with other learned societies. A lighter aspect of official business is seen in a letter written to him by Sir Edward Sabine (1788-1883), Treasurer of the Society, about C. R. Weld, an assistant secretary, who wrote a scholarly *History of the Royal Society* in two volumes. It became known that Weld had "introduced a lady into his apartments in Burlington House where he lived. In my last interview with Weld he seemed anxious to explain that the lady in question is a lady of independent fortune". Not surprisingly, Sabine thought that irrelevant and Weld was to be dismissed after due notice.

Failing eyesight led to Sharpey's retirement in 1872 but he was elected to serve on the Council for a further two years. Of his years in office

Thomson said of him "All those who attended the Royal Society or took part in its proceedings are well aware of the strong and steady interest which Dr. Sharpey took in all its affairs, and of the great amount of anxious care and judicious labour which he devoted to the promotion of its welfare".

## THE GENERAL MEDICAL COUNCIL

Following a long period of debate about the reform of medical practice the Medical Act of 1858 brought into being the General Council for Medical Education and Registration, usually known as the General Medical Council. To the public it is best known as a regulatory body with powers to strike off members who do not meet its standards of competence or ethics. Its educational role, to lay down the subjects and the standards to be attained, was not so much in the public eye but obviously was of great concern to the medical schools and the professional bodies. The Council had a wide membership of elected representatives of the universities and Royal Colleges together with a number appointed by the Privy Council to which it reported (not to Parliament). Sharpey was appointed by the Crown in 1861; his friend Allen Thomson joined him as the representative of Glasgow and St. Andrew's Universities. The normal term of office was five years but he was re-appointed in 1866 and again in 1871; he served for much of that time on the executive committee and he became chairman of the finance committee, a reflection of his standing with fellow councillors. Although he was an active member of the Council, his contributions went largely unrecorded. His main work was as a member of the education committee and he, together with Parkes (1819-1876) of University College and Syme of Edinburgh, were responsible for the changes in the medical curriculum which led to the division into pre-clinical and clinical stages. It was recommended that in all the pre-clinical subjects, not just anatomy, there should be a greater emphasis upon practical work and this led to the establishment of university laboratories and the employment of specialist staff who did not necessarily have medical qualifications. One of the first of this new breed of medical school teacher at University College was F. J. M. Page (1848-1907) who graduated BSc from the Royal School of Mines and then worked on electro-physiology with Burdon Sanderson from 1875 to 1885.

This low-key activity at the General Medical Council would have been in keeping with Sharpey's temperament, untroubled by dramas such as the Cooper-Syme affair at University College or the Lee controversy at the Royal Society. That he was twice re-appointed to the Council may be taken as an indication of the confidence placed in him by his peers.

## THE ROYAL COMMISSION ON SCIENTIFIC INSTRUCTION AND THE ADVANCEMENT OF SCIENCE (THE DEVONSHIRE COMMISSION)

This body was established by the Government in 1870 as a response to the growing awareness of the need for Government intervention and financial support for the science based industries upon which Britain's wealth depended. Thomas Huxley was already a powerful influence in the cause of scientific education and the relatively unknown Alexander Strange FRS (1818-1876) had suggested the need for a Ministry of Science; it was largely due to their advocacy that the Commission was appointed, under the Chairmanship of the Duke of Devonshire, to consider the relations between government and science. Sharpey, together with his friend Huxley were two of the eight appointed Commissioners. Sharpey assiduously attended the sittings and often took the chair but he did not himself present any evidence or became personally identified with any of its main recommendations. He was, as ever, diligent, conscientious and self-effacing but clear-sighted in questioning the witnesses called before him.

Eight reports were issued between 1872 and 1875; the titles of them give some idea of the very wide scope of the inquiry and of the work that Sharpey had to comprehend in the years immediately preceding his retirement. The first report dealt with the London higher technical colleges established at South Kensington after the 1851 Exhibition had given impetus to the development of this area for science museums and colleges. As a result of the Report the Royal School of Mines and the Royal College of Chemistry were merged and subsequently became Imperial College of Science and Technology within the University of London.

The second Report considered technical education at the level of the mechanics institutes and similar bodies in the rest of the country. For a number of years Huxley and Michael Foster had advised teachers on the art of science teaching.

The third Report concentrated upon science teaching at Oxford and Cambridge stressing the importance of some knowledge of literary culture among those who wished to concentrate on science. Both universities had fallen behind in the provision of museums, libraries, laboratories and scientific staff and the Commissioners made several recommendations for improvements one of which was the creation of a chair of physiology at Oxford, later to be established with Burdon Sanderson, a Sharpey man, as its first professor. Their declaration of the importance of combining teaching and research included a passage which could have been written by Sharpey himself since it epitomised so exactly his own philosophy of a university: "If once a teacher ceases to be a learner, it is difficult for him to maintain any freshness of interest in the subject which he has to teach and nothing is so likely to awaken the love of scientific enquiry in the mind of a student as the example of a teacher who shows his value for know-ledge by making the advancement of it the principal business of his life".

The fourth Report dealt with museums and public instruction one outcome of which was the establishment of the Science Museum in London.

The fifth Report considered science teaching in London, at University College and Kings College, and in other university colleges in England. In particular Manchester was praised for its distinction in science and engineering which the Commissioners thought deserved State aid.

The sixth Report dealt with science teaching at public and endowed schools which they found to be very unsatisfactory, concluding that there was an almost total exclusion of science from the teaching of the upper and middle classes which amounted to a national misfortune.

The seventh Report, on science at the Scottish universities, brought forth fewer criticisms since education in Scotland was more under local control and not so influenced by central government.

The eighth and final Report came back to the central issue of the relation between the Government and science. It covered every subject from astrophysics to zoology, finding where there was an important public interest which it was the duty of the State to support. Many of its proposals could not be accepted at the time since the prevailing philosophy was for minimal government involvement in the lives of citizens, an attitude which has now been radically changed as health, wealth and security have increasingly become science-based activities.

ANIMAL EXPERIMENTS AND THE LAW

Concern about the ethics of performing experiments on living animals had been expressed from before Sharpey's time as a medical scientist. The Royal Society for the Prevention of Cruely to Animals had been founded in 1824 with support from the upper classes who sought to stop the unthinking cruelty often found in agriculture and horse transport and the plebeian pursuits of cock-fighting and bull-baiting. It published a pamphlet on experiments in 1827 but it did not become seriously involved with the anti-vivisection debate until the 1870's when it became very influential in moulding opinion. In 1824 Magendie had performed some experiments in London which, in the days before anaesthetics, aroused considerable anger in Parliament and the press by the infliction of severe pain upon dogs with little or no scientific justification. Although he died in 1855, his reputation as a callous and uncaring investigator smeared the name of physiologists for years to come.

The agitation for some form of control of animal experiments took flight in 1870, the year of Burdon Sanderson's appointment to University College, of Foster to Cambridge and of the Royal College of Surgeons requirement for courses in practical physiology. There were many articles in newspapers and journals with strong views being expressed by men of influence on both sides. That this pressure was heeded by scientists was shown by the formation, in 1870, by the British Association of a committee to draw up guide lines for animal experiments. Burdon Sanderson and Foster were among its ten members and a report was issued at the Edinburgh meeting of the Association in 1871 which recognised the need for some form of regulation.

In 1873, the publication of the *Handbook for the Physiological Laboratory* evoked a widespread reaction among the intellectual public since it gave, unintentionally, the impression that no anaesthetics were used and that such experiments were widely performed by students. Although Burdon Sanderson was later to correct this view, nevertheless it did an enormous amount of damage to the image of the physiologist.

An event which acted as a trigger to launch a new wave of anti-vivisection activity was the demonstration in 1875 by Eugene Magnan, from France, to a meeting of the British Medical Association in Norwich, of the induction of epilepsy in dogs by the injection of absinthe. No doubt the previous demonstrations by another Frenchman, Magendie, were recalled. The agitation was such that the RSPCA sought to prosecute Magnan and three doctors from Norwich for cruelty under the Martin Act of 1835 which made wanton cruelty a criminal offence. Magnan could not be brought from France to face prosecution; the three doctors were found not to have been criminally involved and so the case was lost but it was clear that something had to be done to assuage public opinion, on the one hand, and to protect the emerging science of physiology in Britain on the other.

The articulate group of anti-vivisectionists brought forward a plan of regulation which was presented as a Bill to the House of Lords by Lord Henniker in May 1875.

At the same time the physiologists and their supporters, including Darwin and Huxley, drafted their views on regulation and had them presented to the House of Commons by Lyon Playfair, the scientist MP for Edinburgh University. Their concern for the future of physiology was reflected in a letter sent to Playfair by Sir Edward Frankland FRS (1825-1899), a chemist with no direct interest in physiology, who wrote:

> "Physiological and biological research of all kinds is at a scandalously low ebb in this country. It is no exaggeration to say that more of such researches are made in the laboratories of Germany in one academical session than in those of the United Kingdom in several years. Experimental research in some branches of science, chemistry for instance, may possibly remunerate the discoverer, but in physiology this can never be the case. The physiological investigator is the most disinterested worker and he requires and deserves all the encouragement which the state can afford…The incentives to physiological research in this country are at present so feeble as to cause this most important branch of experimental science to be almost entirely neglected".

Owing to the timing there was no possibility of either Bill being passed in that session of Parliament but the Government was aware that the issue would not disappear and so it announced the formation of a Royal Commission on the Practice of Subjecting Live Animals to Experiments for Scientific Purposes, to Investigate and Make Recommendations. The title reflects a neutral position; it was not an anti-vivisection commission although this was its abbreviated name. It sat from July to December 1875 and questioned 53 witnesses, one of whom was Sharpey who gave his evidence on the 6th July.

His understanding of contemporary physiology was shown in his opening statement that the application of physics to the phenomena of life, particularly in making exact quantitative determinations, was one of the characteristics of modern physiology.

He said that he had carried out experiments on animals and believed that they were essential for the science of physiology as shown by the classical work of Harvey, Hales, Poiseuille, Ludwig and others including Magendie. He was convinced of the value of anaesthetics and thought that there were very few occasions when their use might nullify an experiment. Asked about Magendie's work he replied:

> "I may mention to the Commission that when I was a very young man studying in Paris, I went to the first of a series of lectures which Magendie gave upon experimental physiology, and I was so utterly repelled by what I witnessed that I never went back again. My objection in these experiments was two-fold. In the first place they were painful….and they were made without any sufficient object. As an example Magendie made incisions into the skin of rabbits to show that the skin is sensitive. Now surely all the world knows the skin is sensitive, no experiment painful or without pain is needed to prove that".

Questioned again he agreed that Magendie had added valuable results to physiological knowledge but he spoke about the infamous experiment whereby, in the absence of anaesthetics, Magendie had substituted the bladder for the stomach in order to show that external pressure, rather than the stomach muscles, could produce vomiting; "surely", Sharpey comments, "we did not need that experiment to show such a simple result".

He was questioned about the *Handbook* which had been dedicated to him: "but that does not mean that I adopt all the conclusions in it" he replied. He himself would not have demonstrated all the experiments in the book to a class. He agreed that curare could never be used as a sole anaesthetic although it was very useful when used in conjunction with one. He referred

to artificial ventilation by means of a pump, as used by Ludwig;" I have
seen it, there is a similar one at University College, it is a gas engine".

The question of performing experiments on man was raised and Sharpey
gave several examples from history but when asked about the propriety of
carrying out painful experiments on convicted criminals his humanity spoke
out strongly against them: "the penalty of law may be death but not the pain
that would be incurred previously; even if the procedure was not severe
and the offence not a capital one there might be a danger in it".

He did not favour regulation by law, preferring to rely upon the integrity
of scientists, but if legislation was thought necessary he would not oppose
it. Finally, he quoted a passage from Haller," one of the greatest physiologists of all time and author of perhaps the grandest work on physiology that
ever appeared and who was a great vivisector:

> "But it is not sufficient to make the dissections of the dead bodies
> of animals. It is necessary to incise them in the living state. There
> is no action in the dead body; all movement must be studied in
> the living animal, and the whole of physiology turns on the
> motions, external and internal, of the living body. Hence no
> progress can be made in investigating the circulation of the blood
> and its more recondite movements, or the respiration, or the
> growth of the body and the bones, the course of the chyle, or
> the motions of the intestine, without the sacrifice of living things.
> A single experiment will sometimes refute the laborious spec-
> ulation of years".

He concluded with a quotation from Haller in the original Latin which
made a scholarly end to his evidence.

A Bill was brought to Parliament by Lord Carnarvon early in 1876
which raised the fears of many of the active physiologists that it would be
unduly restrictive especially in that it would have banned all experiments
on cats and dogs. Sharpey was in touch with Burdon Sanderson about the
draft Bill and he later wrote to Schafer about it:

> "Lord Carnarvon's Bill is in many respects very objectionable
> - but it is not so bad in essence as in form. If you study it atten-
> tively you will see that it takes the shape of a strictly prohibit-
> ing bill - but then by means of a proviso, one may 'certify'
> himself or at least get certified out of its most oppressive enact-
> ments.... The clause requiring that the experiment must be

'absolutely necessary for the advancement of medicine and the alleviation of human suffering' would interfere with schools such as Cambridge and Oxford - where the studies are not directly medical. Again the proposed bill would allow any restless associate or individual to harass physiologists by vexatious prosecutions - who for all we know would not be slow to engage in such. There must be some protection against this, such as requiring the concurrence of the Attorney-General in instituting any prosecution etc".

After much discussion and lobbying by all sides, a number of amendments were agreed and the Royal Assent was given to the Cruelty of Animals Act on the 15th August 1876.

Its provisions were less onerous than had been feared by the scientists and were far more liberal than had been hoped for by the anti-vivisectionists who called it "a physiologists charter". It now became possible, subject to holding the correct certificate, to use cats and dogs, and the Act protected licence holders from prosecution by members of the public; it was in the hands of the Home Secretary to prosecute for any infringements of the law. This was a point of great importance to working physiologists and it appears to have been included at the suggestion of Sharpey. In this and in his evidence to the Committee his reasonable, scholarly and humane attitude must have made a strong impression.

He was never a licence holder himself, having retired two years before the passing of the act, but he would have been amused to find his name listed in *The Vivisectors Directory* published in 1884 by one of the London societies. It contained the names and qualifications of men, in Britain and abroad, who had any connection at all with animal experiments. There were suitably graphic quotations selectively taken from their publications or as in Sharpey's case, from their evidence to the Commission. He would have been at home in the company of many of his former pupils and colleagues. For a full discussion of the anti-vivisection movement the masterly book of R. D. French (1975) should be consulted.

## THE PHYSIOLOGICAL SOCIETY

It may appear surprising to find how little connection there was between Sharpey and the Physiological Society but it must be remembered that by the time it was founded, in 1876, he had retired from both the Royal Society and the University and was quietly cataloguing the books he was to donate to University College library.

The main stimulus to the founding of the Society was the threat of legislation by the Government in response to the Report of the Royal Commission which was published in January 1876. The growing group of physiologist saw the need to establish some degree of public trust in them and the importance of speaking with one voice on behalf of their developing profession. It would have done no good to their cause by appearing arrogant or defensive; many of those who supported the anti-vivisection movement were men and women of social and intellectual standing.

The formation of a society which was able to regulate physiological experiments had been suggested by Marshall Hall as long ago as 1831 in his work on *The Circulation of the Blood*. He was a humane investigator who proposed certain basic principles for the conduct of experiments on animals which would be observed through the formation of a "Society for physiological research". He raised the question again in 1847 saying "We are greatly in need of a code of medical ethics. I have ventured to sketch what appears to me to be ethical in regard to experiments in physiology". This led to an attack by an anonymous correspondent to which Hall replied at length re-stating his view that "I would again urge upon the profession the institution of a Physiological Society, the objects of which should be, to deliberate on proposed experiments to adopt or reject them, to assist in their performance, to witness their results, to render their repetition, as far as possible, unnecessary, and thus really to prevent cruelty, to promote physiological science, and, as has now become requisite, to protect individual character." This far-seeing proposal anticipated the ethical committees which we have today. By coincidence, his paper appeared in the *Lancet* barely a month after Liston's first use of ether in surgery and in an issue which described one of the first ether vapourisers, but he was just too early to consider the implications of anaesthesia in experimental physiology.

About twenty years later, with the increasing threat of Government legislation, it was recognised that physiologists should unite in their own defence. The first step towards the establishment of the present Physiological Society was the calling of a meeting on the 31st March 1876 at Burdon Sanderson's house in Queen Anne Street. He took the chair and there were eighteen others present including Sharpey, Foster, Marshall and Schafer. At further meetings on the26th April and the 5th May, which Sharpey did not attend, a constitution was agreed and a list of possible members drawn up. The inaugural meeting of the new Physiological Society was held on the 26th May with Foster in the chair and twenty one other members. Sharpey was not there but he, and Charles Darwin were elected the first Honorary Members.

The Society was more than merely a defence organisation; it was, and remains, a learned society dedicated to the advance of physiology; in its early days it was organised along the lines of a gentleman's dining club at which papers were read in a convivial environment. This would certainly have suited Sharpey; he attended five meetings between March 1877 and January 1878 but he did not give any papers. At the meeting in December 1877 he was placed on the committee which had been set up to consider the working of the 1876 Act and so was involved with the Society business. At the same meeting his guest was Allen Thomson who joined the Society in the next year and eventually he too was elected an Honorary Member.

Had the Society been founded ten years earlier Sharpey would have undoubtedly been more actively involved with it but he must have experienced considerable satisfaction to see this new sign of maturity in a subject which had been the main interest of his life for forty years.

## THE HUNTERIAN MUSEUM

In December 1799 Parliament agreed to the purchase of the collection of anatomical specimens collected by John Hunter (1728-1793) during his long career as a comparative anatomist at the Royal College of Surgeons. Responsibility for this magnificent collection was vested in a Board of Trustees, some *ex officio* and some by appointment; among the latter was Sharpey who served from 1862 until his death. It must have been a very congenial appointment since he had shown an interest in the subject from his work on cilia in the 1830's and it is a measure of his enthusiasm that he had the fourth highest attendance at meetings of the Trustees during his nineteen years in office. The vacancy caused by his death was filled, most appropriately, by his friend Allen Thomson.

## THE ROYAL SOCIETY OF MEDICINE

Sharpey was elected as a Fellow of what was then called the Royal Medical and Chirurgical Society in January 1837, soon after his appointment to University College; one of his sponsors was Robert Lee with whom he was then on good terms. The Society always had a strong connection with the scientific aspects of medicine, which would have appealed to Sharpey and he served on the Council between 1848 and 1850 but he did not publish any papers in its Proceedings and after his appointment as Secretary of the Royal Society in 1853 he took no further part in its affairs.

## THE ROYAL INSTITUTION

An office held by many medical scientists is that of Fullerian Professor of Physiology at the Royal Institution. It was founded under the terms of a Trust made by John Fuller in 1834 which provided funds for professors of chemistry and physiology with a tenure of three years. It is often held as an additional post conferring an honour, a small salary and the opportunity to communicate with a wider audience than the University.

Sharpey was never appointed to it; whether he was offered it is not known but during his lifetime it was held by a number of his colleagues and former pupils (Huxley, Marshall, Foster and Schafer) and so he can be credited with a measure of reflected glory.

## THE MICROSCOPICAL SOCIETY OF LONDON

This Society, now the Royal Microscopical Society, was founded in 1839 and Sharpey was one of the founder members but he does not appear to have taken much part in its activities and his name is not on the list of members after 1849. This might seem at odds with his interest in promoting the use of the microscope by his students but he was becoming involved with the Royal Society as a member of Council from 1844 and then as Secretary from 1853 which diverted him from other activities. In his London days he was never a serious student of histology as a field of research; except for Sharpey's fibres, he did not make any original discoveries or devise any new techniques.

CHAPTER 4

# Sharpey – The Professor

## THE WRITER

SHARPEY WAS BY no means a prolific writer, rather the opposite, and he did not undertake what he certainly had the talent to do, namely write a *magnum opus* on anatomy or physiology; he came nearest to this through his joint editorship of Quain's *Elements of Anatomy* which he took through several editions. He did, however, produce a small number of original papers as well as the various reviews and printed lectures derived from his position as professor. His publications may be divided into those written before and those written after his move to London; it is the former which brought him recognition as a scientist upon the basis of which he obtained his post at University College. In the following survey of his writings the numbers refer to the list in the bibliography (Note 4).

His thesis for the MD degree at Edinburgh, written in Latin, as was then the custom, was entitled *De ventriculi carcinomate* (1) and was dedicated to his step-father William Arrott and to Robert Knox. It was read to the Royal Medical Society of Edinburgh in 1823, presumably in English, and published, posthumously, in 1892 as *On cancer of the Stomach,* in a collection of similar dissertations. In it he described the symptoms, morbid appearance, diagnosis, prognosis and treatment of the disease. He noted that there were no cures, blood letting he regarded as useless and the only treatment was by palliatives such as belladonna and moderate eating to retard its growth, and finally doses of opium to alleviate pain. It reads as a workmanlike account of contemporary knowledge but without any new findings.

His next professional publication was his Thesis for his Fellowship of the Royal College of Surgeons of Edinburgh. This was a study of so-called false joints (2) and was dedicated to James Syme. Sometimes a broken bone does not repair with a hard, unyielding callus but a degree of movement persists between the two parts producing lameness; the fracture is un-united creating a false or artificial joint. He gave a detailed account of the subject with many references in French, German and Latin and concluded that it was better to live with the disability rather than risk having the joint broken and re-united at a second operation, no small undertaking in those pre-anaesthetic days.

In the same year (1830) he wrote an account of what became his sole claim to scientific originality, the discovery of cilia (3), which receives a citation in Garrison & Morton's *Medical Bibliography*. In the course of work on the development of the frog's egg, he observed that the surface of the immature tadpole "possessed the power of exciting currents in the water contiguous to it in a constant and determinate direction" and "on examining a portion of the gills with a powerful lens I perceived that it was beset with minute cilia, which are evidently instrumental in producing the different currents". He saw the same movement in the gills of the common mussel but not in the chick embryo. Although limited in its scope, it was a well-written paper which has a permanent place in the history of the subject.

Other topics occupied him until 1835 when he was stimulated to renew his interest in cilia by the appearance of the work of Purkinje and Valentin in Muller's Archiv for 1834 in which they described ciliary movement in the birds oviduct and the mammalian Fallopian tubes and respiratory system. He wrote (7) a translation of their work and added some further observations of his own. From his extensive reading he found that several previous authors, including his future colleague Grant, had commented upon the vibratory fluid movements produced by cilia, without actually identifying them. He was never one to chase claims for priority and he was content to have witnessed what he called "one of the most beautiful phenomena in nature".

His chapter on Cilia in Todds *Cyclopaedia of Anatomy and Physiology* (8) was a scholarly review which is still worth reading today. He considered the animal kingdom systematically giving references to the existence of ciliary motion which he had found in an extensive search of the literature. He frequently provided first-hand descriptions of his own, unpublished, observations on cilia in the sea anemone, the sea urchin and the sea mouse and mentions experiments in which he used a suspension of charcoal to determine the direction of movement of the ciliary current in the nose of the rabbit and the trachea of the dog. It is a remarkable piece of work for someone who, at the time, was very busy establishing his school of anatomy.

In his remarks on a supposed spontaneous motion of blood (3) he pointed out that ciliary action would account for the movements observed by previous authors but he added nothing new of his own.

His paper on *Infusoria* (5) is an example of his dedication to the microscope and of his familiarity with European work then being undertaken. Christian Gottfried Ehrenberg (1795-1876) was Professor of Medicine in Berlin where Sharpey had met him. He was an expert microscopist who described the organelles of protozoa and bacteria, which he lumped together as infusoria, and he published a beautifully illustrated work on them in 1838. Sharpey's note brought Ehrenberg's important studies to the notice of English readers.

His paper on the blood vessels of the porpoise (6) is a more conventional piece of gross anatomy in which he described some of the special features associated with diving mammals. He mentioned the convoluted arterial plexuses in the fin and the neck; the narrowing of the internal carotid artery where it enters the skull; and the plexiform arrangement of the thoracic aorta known as the *rete mirabile*. It was not a particularly original study and only an abstract was presented.

His last contribution before moving to London was his account of the *Echinodermata* in Todd's *Cyclopaedia* (9). It shows the same erudition as before and includes some original observations on the vascular system of starfish. His reference list includes papers in German, French, Italian and Latin and he seems so thoroughly at home with the subject that it is surprising he did not become a comparative anatomist.

On the 3rd October 1836 Sharpey gave his inaugural lecture at University College; it was printed in the Lancet (10). Aware of the controversy surrounding his appointment, he makes a frank appeal for tolerance and understanding: "I come among you as a total stranger, and am aware that I have yet to gain your confidence". He lavishes praise upon his predecessor, Jones Quain, and entreats his audience to extend to him their "indulgent consideration" adding "I feel that until I have earned your confidence and friendship, I shall stand in need of your favour". He based the rest of his lecture on the nature of physiology and anatomy pointing out how Harvey progressed from the anatomy of the valves in the heart and veins to the functional experiments by which demonstrated the circulation of the blood. As an encouragement to others and a reflection of his own modest attainments, he went on to say "Discoveries of such magnitude can fall to the lot of very few; but there are lesser jewels in the mine of science which he who pursues his studies with a philosophic spirit can ever hope to discover".

His interest in function led him to pose the question, well ahead of his time, as to how glands produce different secretions, eg. tears, bile, urine, when starting from the same common material, the blood. He made a strong case for the inclusion of comparative anatomy in his course as an exercise in scholarship: "I need not enlarge on the intrinsic merits of this engaging study, on the endless variety of objects which it affords to stimulate rational curiosity, or the profound and extended views of organised nature which it presents to the philosophic inquirer". This is the broad, intellectual approach to medicine which the founders of the new university looked for and explained why the appointment committee preferred Sharpey to Grainger.

He pointed out that other animals, although of the same general design as man, frequently have some features exaggerated; "It is obvious that in such cases the investigation of these parts in man must be greatly facilitated by previously examining them in other animals where their texture is unravelled and magnified to us by nature, and where we are enabled, as it were, to read the same truth in a larger type". This is an eloquent statement of the value of comparative studies; a modern example of which would be the use of the giant nerve fibre of the squid in the study of nerve impulses. Continuing this theme, Sharpey described the venous reservoirs in the seal and other diving animals, a subject he had studied in Edinburgh, and he called attention to the value of embryology in understanding how complex organs, such as the heart or brain, developed from simple structures. Finally he urges upon students the value of practical work, of knowing the physical reality of what he learns about in books, by the intelligent use of the microscope. It must be remembered that in 1836 there was little else except microscopy as an investigative technique and it was not until thirty years later that the scope of practical classes extended to include animal preparations.

This lecture by Sharpey was far-sighted, erudite and stimulating; the same can not be said of a later series of five introductory lectures published in the *Lancet* (11). He might have been expected by 1840 to have established his credentials as a physiologist but in this respect they are a great disappointment. The style is verbose and the content mundane. He spoke of the differences between living and non-living matter and between animals and plants and outlined the principles of comparative anatomy but nowhere did he mention Harvey, Hales or Mayow. Perhaps the purpose of these lectures was to provide a background for students with no previous knowledge of biology before going on to give a more orthodox course of physiology during the term. He gave advice about text-books: on general anatomy Quain was essential and a few others, including one by his friend Grainger; on physiology he strongly recommended the new English translation of Muller which he knew students found too difficult but which was the most up-to-date. Others he mentioned were those by W.B. Carpenter, James Bostock and Elliotson's translation of Blumenbach. His own favourite, which he admitted was out-of-date and too ponderous for ordinary reading, was the eight volume work by Haller; it was, he said, still a book he found indispensable.

In 1842 Baly's translation of Muller's *Elements of Physiology* was published. It contained an eight page supplement by Sharpey on the structure of decidua (12) which, as would be expected, provided many historical references and a well-ordered account of the subject which included many of his own dissections.

His work as an editor of anatomical text-books started about the this time (13) and continued from 1848 until into his retirement with successive editions of *Elements of Anatomy* (14-17) in which he included for the first time a description of Sharpey's fibres. This, his only eponymous attribution, justifies some further explanation.

Careful examination of the underside of the periosteal membrane of bone reveals a network of perpendicular fibres running deeply into the bony lamellae, but not into the Haversian systems, apparently bolting them together and to the overlying periosteum. From the manner of their insertion they are called perforating fibres. They may be frequently seen on the flat surface of detached lamellae projecting like nails driven through a board. They are composed mainly of collagen, which may become calcified, but some are of elastin and these may shrink or be lost in dried bone thus leaving open channels.

Sharpey's original preparation was drawn for him by Allen Thomson and appeared as a plate in his section on bone in the 5th edition of *Elements of Anatomy* (1848), the first time he was a joint author of this work. He described the perforating fibres but without announcing any new discovery or claiming any priority. However, in the 7th edition he did add the phrase "first noticed by me" when describing the fibres. This was withdrawn from the 8th edition but by then a new eponym had been created; henceforth they became known as Sharpey's fibres, a term now widely accepted. He must have observed them before 1840 since a student, E. Ballard, mentioned them in his notes on Sharpey's lectures.

According to Schafer in his *History of the Physiological Society*, "It is to Kolliker that we owe the term Sharpey's fibres to the perforating fibres of bone which Sharpey described" but exactly when and where this attribution was made is not clear.

Rudolph Albrecht Kolliker (1817-1905) was a life-long friend of Sharpey and was often mentioned in letters to Allen Thomson. He was a frequent visitor to Britain where he sometimes stayed with Sharpey in London or was in his company at scientific meetings such as that of the British Association in Glasgow in 1855. He was Professor of Microscopic Anatomy at Wurzburg, Bavaria, from 1847 to 1905. In 1860 he was elected a Foreign Member of the Royal Society and in 1897 he was awarded the Copley Medal. Like Sharpey, he was principally a microscopist but he had a wide range of interests which included embryology, marine biology and experimental physiology. He was the first to show, in 1856, by the use of the rheoscopic nerve-muscle preparation of the frog, that the cardiac impulse preceded the contraction of the heart. He made a systematic study of human histology and showed the presence of muscle fibres in blood vessel walls and the continuity that exists between a nerve fibre and its cell body, both findings of fundamental importance. Arising from this original work was one of the first text-books of histology, written in German in 1850 and published in England by the Sydenham Society in1853. At the meeting of the British Association in Glasgow, Kolliker agreed to produce a new edition of his book to be translated by George Busk, of the

University of London, and T. H. Huxley. Now virtually a new work, this was published in 1860 under the title *A Manual of Human Microscopic Anatomy* and its frontispiece carried the inscription "To his dear friend Professor Sharpey Sec. R. S. in grateful remembrance of much kindness and encouragement and in token of high and sincere esteem, this work is dedicated by the author". Strangely, it contained no specific reference to Sharpey's fibres although he refers to the 6th edition of Quain (1856) in which "the histological part (is) done by Sharpey most excellently". The origin of the eponym thus remains uncertain.

No doubt Sharpey was pleased with his eponymous distinction but he was never one to seek recognition for all that he did and when his priority was challenged he at once gave way graciously. The challenge, which was by no means strongly argued, came in 1875 in a paper in a little known journal by Clementi who asserted that two of his countrymen, Gagliardi in 1689 and Troja in 1775, had a prior claim for the discovery of the perforating fibres in bone. Sharpey had a formidable knowledge of the European literature of his subject and for him not to known of these papers is a measure of their obscurity. Nevertheless, he examined their work carefully and agreed that, to some extent, he had been anticipated by Troja. In a postscript to a later paper by Schafer in 1878 (20) he concluded with wry good humour: "that when I first observed these fibres I had no idea that they had been recognised before, still less did I imagine that the subject of my observation would ever acquire such importance as to lead to a formal claim of priority on the part of Italian science". And the anatomical world has continued to call them Sharpey's fibres ever since.

The address in physiology which he gave to the British Medical Association in 1862 (18) is the nearest he came to what we would recognise as physiology today. He spoke about the progress of physiology over the last 25 years, the time he had held his chair at University College, and of the way vital activities were increasingly being explained in terms of physics and chemistry. As an aside he remarked that physiology brought no material rewards to its votaries, only the satisfaction of being devoted to science and achieving honourable distinction. He commented upon the increasing and often ignorant, opposition to experiments on animals and, while not condoning cruelty, defended their use in pursuit of medical knowledge. Strangely, he did not mention anaesthetics, which have made animal experiments so much more acceptable.

He then turned to the new techniques which have enlarged the scope of physiological investigations. Microscopy had been improved enormously by the achromatic lenses of J. J. Lister; the ophthalmoscope of Helmholz, Czermak's laryngoscope, and the thermopile and the galvanic multiplier of du Bois Reymond which made possible a deeper examination of nerve and muscle. He referred to Poiseuille's manometer which allowed blood pressure and other variables to be recorded on a revolving clockwork cylinder (the word kymograph had not come into use). The speed of nervous conduction (Helmholtz), liver glycogen (Bernard), oxygen transport in the blood (Lothar Meyer and Harley), Wallerian degeneration, nervous control of the digestive glands (Bernard) and vasomotor nerves (Bernard and Waller) were all subjects which had been discovered in his time and with which he was familiar.

A few years later he delivered a presidential address to the British Association in Dundee (19) but he did not elaborate to any extent upon recent advances in physiology. He mentioned Burdon Sanderson's work on the influence of respiration on the circulation by the use of the sphygmograph and the experiments of Fick and Wislicenus in Switzerland and of Frankland and Parkes in England which proved that muscular activity was not brought about through the metabolism of muscular tissue, as measured by the loss of nitrogen, but required carbohydrates as a fuel. It says much for his awareness of scientific advances that these findings had been published only the previous year. As became a now elder statesman of his science, he commented upon its growth during his lifetime and upon the need for journals and societies to inform the public of the progress of knowledge.

His last paper (21) was the reworking, in his retirement, of some clinical observations he had made in 1824. They concerned a woman patient who suffered from progressive 'torpor' with a loss or reduction of most bodily functions. She did not respond to blistering, electrical shocks or drugs but gradually, over a period of months, made a complete recovery without, however, having any memory of the events. Sharpey considered this to be some form of re-education of the brain although he expressed no clear view about what this meant. It is a well-written fragment from a man of 77 who could still take an interest in the world around him.

Sharpey wrote clearly and usually concisely; his article on Cilia was a model of good scientific literacy. But he was not a prolific writer, his longest work was in Quain's Elements of Anatomy but it is not clear how much of that was from his own hand except for the section on histology. His was perhaps more attuned to daily tasks which could be completed quickly, such as he would encounter as an administrator at the Royal Society, rather than to the laborious work of a scholar with its need for extensive reading, summarising, organising and then writing an original treatise or a major text-book. It has been noted that, unlike his colleagues Allen Thomson and John Marshall, he never wrote even a minor text-book of physiology although he undoubtedly had a very wide knowledge of the subject from earliest to modern times.


## THE EXPERIMENTER

Sharpey was never a dedicated research worker but he did undertake experiments from time to time in order to satisfy himself that something was correct or to demonstrate a point to his pupils. He certainly had 'hands-on' experience at the laboratory bench and it is the aim of this section to examine to what extent he was a practical physiologist.

According to Schafer, Sharpey was a great teacher but not really a physiologist. "Of the physiology he taught very little was acquired as a result of personal investigation and his knowledge of methods was nil. But he had a clear ideas regarding the principles of the science and an extraordinary facility for imparting his ideas and for interesting his hearers in them". This is perhaps an underestimate of Sharpey's practical abilities since there are many records of experiments by him which involved the use of scientific instruments.

Writing to Thomson from Edinburgh in 1836 about repeating Poiseuille's experiment on a dog he states: "the pipes were put into the carotid and crural and the thing did well, the perfect equality of pressure is no exaggeration, it was quite conspicuous except when, once or twice, the tubes got slightly obstructed….I suggested to Christison that we should make trial with it of the force of the heart under the influence of different poisons etc.". This shows him to be capable of using a modern technique and obtaining confirmation of an important finding; it is a pity that he was not led on to make original investigations in the same field, he would have been an early pioneer.

In London he continued to demonstrate his intellectual curiosity by undertaking experiments on various topics. As early as 1837, he told Thomson: "I tried Humboldt and Muller's experiment on frogs muscles and succeeded perfectly"; this refers to the use of electro-physiological techniques which were then just emerging. Muller, in his text-book in 1834, had confirmed that the Bell-Magendie law applied to the frog.

Writing in 1847, Sharpey wrote:

> "I have been trying an experiment of Weber intended to show a different effect from stimulating a nerve and a nerve centre respectively. A current or rather a series of interrupted currents from an Electro-Magnetic Rotatory apparatus is sent through the spinal cord of a frog - a tetanic state of the body is produced & this *continues* a little time (a minute or two in my experiments) after the application of electricity ceases - but according to Weber & Volkmann when the same state is caused by passing electricity through a *mere nerve* it ceases *the moment* the electricity is discontinued.
>
> I do not get the same result - I find the tetanic state persists some time after the electricity is stopped in *both cases* - it may endure a little longer via the cord but there is no absolute difference....I wish you would repeat the experiment for I saw Todd the other day who told me he got the same result as the German Physiologists - I had not then tried it - I used strong large frogs".

Here again his enquiring mind leads him to confirm for himself results which he could not accept uncritically.

In 1853 he was present at an important experiment performed by his friend A. V. Waller at University College. Although he was there only as an observer, he had provided the facilities for Waller and he must have had an informed interest in the proceedings. This is supported by Lister who recalled, when a student in 1849, "the flood of light thrown upon Dr. Sharpey's physiological class by his demonstration to us of the effects upon the circulation of the division of the sympathetic nerves in the neck". The background of Waller's work is as follows. In December 1851 Claude Bernard reported to the Societé de Biologie his discovery that cutting the cervical sympathetic nerve in the rabbit led to an increase in the temperature of the ear. He believed that this was due to a rise in heat production previously held in check by the nerve. He did not mention vasomotor activity neither did he try to reverse the effects of section by stimulating the cut end of the nerve. This he did in November 1852 when he demonstrated that all the responses to sectioning, including the calibre of the blood vessels, could be made to disappear by periods of galvanism.

Waller had been working in Paris and was aware of Bernard's earlier paper but towards the end of 1852 he had returned to London and while there he decided to show, and to claim priority for showing, that the effects of cervical sympathectomy could be reversed by stimulation of the central end of the severed nerve. With the co-operation of Sharpey, he performed an experiment on the 10th December and on the 9th February he repeated it before witnesses. He wrote an account of his results in a letter to Sharpey which is reproduced here in full since it has not been published previously. It is a well written account of an important experiment and it shows Sharpey's involvement with contemporary physiology.

Arlington Street February 10th 1853
My dear Sir,

I do myself the pleasure of forwarding to you a short notice of the experiment which I performed yesterday in the presence of yourself, Dr. Carpenter and W. Edwards.

The subject was a cat; the sympathetic having been uncovered was carefully separated from the vagus and left in the tissues for upwards of a quarter of an hour. No effect was produced on the pupil, the nictitant membrane, or on the ear. These parts, compared with the opposite side, manifested no difference.

The cervical sympathetic being raised on a probe, its altered state of innervation was evidenced by a slight contraction of the pupil, protrusion of the haw, [an old term for the nictitating membrane] and nearly at the same time slightly increased fullness of the vessels of the ear with augmentation of the temperature of the part. A ligature being then placed on the symp. all the same symptoms were increased. The temperature of the ear, compared with that of the opposite side, and with what it exhibited previously to raising the nerve, was found to shew an increase of about 1' Centig. which persisted without variation.

On applying the poles of the galvanic apparatus to the nerve above the ligature, it was found that besides producing dilatation of the pupil and retraction of the nictitant membrane, it caused the redness of the ear to disappear. This last effect was very manifest soon after the dilatation of the pupil had reached its full extent, and after a minute or two, the ear was as pale as that on the opposite side.

By alternately applying and removing the galvanic irritation it was seen that the alternate emptying and filling of the vessels could be effected "ad libitum", exactly in the same manner as the action of the pupil and the nictitant membrane. This emptying and filling of the vessels was accompanied with a corresponding decrease and increase of temperature of the part, a slow fall and rise of a thermometer placed "a derriere" to the extent of half a degree being observed. In about 30" there was produced a fall of 0.2' Centig. and at the end of a minute the fall was 0.4' Centig. On prolonging the galvanic irritation for several minutes so as to produce dilatation of the pupil, it appeared to the touch that on the opposite side the temperature was lower than on the sound ear.

Two incisions as similar as possible were made in the sound and the injured ears; on the sound side only a faint outline of blood appeared on the edge of the wound, on the other several drops of blood flowed from it. While the blood is flowing from the wound it may be arrested almost immediately by galvanising the symp.

Decr. 10th. The sympathetic was cut on a full grown cat and 9 days after, the same experiment was repeated on the other side. In both eyes the pupils were contracted and the nictitant fold protruded.

Feby. 6th. The left eye was found to have the pupil and haw in the normal state, which was not the case on the right side. It was found that the left ear was not injected, while the right ear was somewhat injected and warmer.

It is evident from the foregoing observations that the vessels of the ear are controlled by the same nervous power as the pupil, to such an extent that from the one we may deduce the state of the other; so that, as we have seen, if the pupil is slightly affected the injection and temperature of the ear is slightly increased; if by the reproduction of the cervical sympathetic the pupil had recovered its functions the injection and increased heat of the ear have likewise disappeared.

Microscopic experiments show us that after section of the cervical sympathetic, made in the lowest parts of the neck, that the part below the section remains normal, and consequently had its centre in the spinal cord; while the part above the section is disorganised up to the first ganglion and this affords strong evidence that all the fibres have a common origin in the spinal cord. It remained still to be ascertained by direct experiment whether the whole of that part of the spinal cord i.e. the cilio-spinal region, likewise affects the contractile fibres of the vessels of the ear and corresponding side of the face.

With many thanks for the kind interest you have manifested in these inquiries, and the facilities you have afforded me for prosecuting them at University College

I remain my dear Sir

Yours very sincerely

A. Waller

Dr. Sharpey FRS etc etc.

Dr. Carpenter was W. B. Carpenter FRS, at that time Professor of Medical Jurisprudence at University College, who later became Registrar of the University from 1856 to 1879. He had held the Fullerian Chair of Physiology at The Royal Institution from 1844-1848 and wrote a textbook of physiology recommended by Sharpey.

W. Edwards may have been (Charles) Worley Edwards, Waller's brother-in-law, a pharmacist living in Bloomsbury, close to University College.

Waller published these and other results in France at the end of February, unaware of Bernard's later paper and of a similar one by Brown-Sequard published in a minor American journal in the previous November. Strictly these earlier papers deprived him of his priority but on comparing the three it is clear that Waller made the most significant contribution. It was a ringing declaration of the widespread physiological importance of vasomotor nerves, stated with the utmost clarity, in marked contrast to the limited role, within the head, ascribed to these nerves by Bernard and Brown-Sequard.

Sharpey must have enjoyed being involved in this important piece of physiological research. He had known Waller for some time and took an interest in his work on the degeneration of nerve. Waller recalls "In March 1850 when I went over the experiments with Dr. Sharpey, I demonstrated to him the effects of section of the lingual nerve in the dog, on the nerves of the papillae fungiformes, and on the ramifications in the mucous membrane of the tongue".

In their text-book *The Physiological Anatomy and Physiology of Man* (1857) Todd and Bowman, when discussing the force of the circulation, refer to "Dr. Sharpey's experiments. A syringe with a haemodynamometer, to show the pressure, was adapted to the thoracic aorta of a dog and the inferior vena cava opened near the diaphragm. Fresh defibrinated bullock's blood was injected with a pressure of 3.5 inches of mercury, and passed through the double capillary system of the intestines and the liver out of the veins with a full stream. The blood was made to traverse the capillary system of the lungs, by a pressure of 1.5 to 2 inches of mercury, so as to flow freely through the pulmonary veins. Allowing one pound for every two inches of mercury, it would thus appear that a pressure of two pounds was sufficient to complete the circulation through the two abdominal capillary systems and of one pound for the pulmonary circulation". Although limited in scope, this was a good example of the use of the quantitative method in physiology which is much to Sharpey's credit.

In 1851 he tells Thomson that he does not yet know what to think of sugar in the liver. "The fact that the liver yields sugar is no doubt true - I tried it last season when I got Bernard's paper & showed the result to the class - but is the sugar *formed* in the liver? That sugar should continue to be produced in Carnivora which have long kept strictly on animal food - is startling at first - still it is not more remarkable than the production of milk containing sugar in a nursing lioness. I wish I knew more chemistry".

He did something about his lack of chemistry by attending classes; writing to Thomson in 1852 he says:

> "Having long felt the difficulty of mastering (I mean rightly understanding & *retaining*) the results of modern physiological chemistry, I have even begun again at the beginning and worked for the last three months as a humble practical pupil (he was then aged fifty) in the Birkbeck Laboratory (of University College) deeming practice the best means of attaining my end. I thought it best to go through the entire procedure of inorganic analysis before tackling the Animal Chemistry and am tolerably expert in qualitative analysis. I shall continue the work as I

can find time which I hope to do for 2 or 3 hours a day during the busy season. I have been greatly pleased with the occupation which I find a most salutary mental exercise".

His colleague Stanley Jevons, in his memoirs written many years later, remembers the occasion: "He attended Dr. Graham's class this year (1852) with all the other students, and since Easter has been working all day in the laboratory, with Dr Williamson (Professor of Chemistry and FRS) telling him how to do the things. You must not complain of me making messes and blow-ups in the cellar if an old chap of sixty begins to learn to do it". Although Jevons misjudged his age (he was 50), it says much for Sharpey's enthusiasm for learning and his lack of any false pride about his status, that he should take up classes with students a generation younger.

It is clear that Sharpey was capable of performing at the laboratory bench as well as at the lecturers rostrum and this must have brought freshness and vitality to his teaching but he never settled to the daily routine of research with its demands upon his time, the need to devise new techniques and the tenacity to repeat and to modify his experiments until they were successful.

## THE TEACHER

The *British Medical Journal*, in its obituary article in 1880, reflected the widely held view when it stated that "Dr. Sharpey was eminent as a teacher, in fact more in this than in any other capacity. For years, he was the greatest teacher of anatomy and physiology in this country, occupying a position here equal to that held by Johannes Muller in Germany. His style of lecturing was clear, and marked throughout by a candid and at the same time careful appreciation of the labours of others in his department. No new fact or statement was put before his pupils without having been first subjected by him to careful examination; and both his examination and his comments must have had a healthy effect on the mental disposition of his hearers".

According to Allen Thomson: " Sharpey never wrote out his lectures, excepting introductory ones, and he delivered them all without any assistance from writing beyond very short jottings on small slips of paper. He made use of diagrams and pictorial illustrations as well as of anatomical preparations and physiological experiments, and he was one of the first to introduce the employment of the microscope for the practical illustration of his lectures".

Writing in 1932, Schafer recalled his impression, as a student, of Sharpey as a teacher:

> "Although Sharpey had not the advantage of a physiological training, he was an extensive reader and had a sound theoretical knowledge of the science. He was well able to impart his knowledge to an audience of students; no one could listen to him without being impressed by the clarity of his exposition and by his methods of explaining how any problem was to be dealt with or how a particular experiment was performed. He himself had no apparatus except his microscope. The kymograph, that invaluable instrument for recording physiological functions, had been invented by Ludwig in 1846 but Sharpey never used one. Instead he would illustrate how a record of blood pressure or of muscle contraction was made, with the aid of what Michael Foster once termed 'his dear old chimney-pot' – a form of headgear which everyone wore in those days, students as well as professors, which he would revolve in his hand to simulate the smoked drum. His knowledge of the history of physiology was – like Sam Weller's knowledge of London – 'extensive and peculiar', and few things help more to make a subject interesting than appropriate references to its history. We one and all had an affectionate reverence for Sharpey which was not approached in the case of any other professor".

The chimney-pot hat, like the circulating microscope table, has become part of Sharpey folklore but, as we have seen above, he was no stranger to the laboratory and he was instrumental in inducing the Faculty of Medicine to institute practical classes well before these became an examination requirement.

Another tribute to Sharpey's powers as a teacher came from Sir Thomas Barlow, then a leading physician at University College Hospital, in an after-dinner speech in 1912. "I think", he said, "of all our teachers, on the College side, Sharpey was *facile princeps*. What a privilege it was to listen to the principles of physiology expounded with such breadth and sanity. We all of us remember how speculative fallacies were torn to pieces and how thoroughly the well-established facts were marshalled before us, and the lacunae of knowledge clearly indicated. What a model was every lecture. If any lecturer could save his pupils from slipshod thinking Sharpey could".

His qualities as a teacher were recognised outside the University as shown by the tribute paid to him by Alexander Bain (1818-1903), a notable Scottish academic whose career had something in common with Sharpey's. From a humble background, and being largely self-taught, he graduated from Aberdeen University in 1840. He, too, chose an academic life which was centred mainly on Aberdeen and, like Sharpey, he spent some time studying abroad. He eventually became Professor of Logic, a post he held for twenty years. His subject was moral philosophy but he moved towards psychology and in 1876 he founded the important journal *Mind* which he edited for sixteen years. He appreciated the importance of understanding the physiological basis of mind and for that purpose he went to London in 1842 to make contact with those most likely to able to help him. He records in his *Autobiography* how he came into contact with Sharpey. He met first his fellow Aberdonian Neil Arnott, with whom he established a lasting friendship, eventually becoming his executor and biographer. At one of Arnott's weekly bachelor dinners for men of note, he was introduced to Sharpey and, as a result, he subsequently attended the latter's lectures on the brain and nervous system. He remarked of Sharpey later that "His exposition contained the most advanced views then held at that time. In particular he gave a resume of the nature of the nerve force, introducing some of Faraday's speculations as illustrated by his electrical researches". Later, he mentions that "On Sharpey's recommendation I got in Paris Longet's *Physiology* as a help for the senses and for the physiological part of psychology generally" and again "I had Dr. Sharpey's assistance in revising the chapter on the nervous system as well as the physiological parts of *The Senses and the Intellect* (1855). Although Bain was never a physiologist, and therefore not to be counted as one of Sharpey's men (see next chapter) this incident shows something of Sharpey's undoubted standing in the academic world and his wholehearted support for younger men starting an academic career. He was as generous with his knowledge as he was with his money; it was always available to those who cared to use it.

To these recollections must be added several sets of manuscript notes taken down by a number of students who attended Sharpey's lectures and which have been preserved [Note 8]. There are ten collections, the earliest of which is by John Phillips Potter, MB 1841 and afterwards a demonstrator in anatomy at University College. His notes were taken in 1836-1837, Sharpey's first year in office, and cover 26 lectures on general anatomy ie. histology, and 28 lectures on the anatomy and physiology of digestion. The notes taken by (Sir) Richard Quain, the physician, were discovered in an antiquarian bookshop by J F Fulton, the medical historian, who donated them to Yale, his old university. He stated that the " full and beautifully written notes give a clear idea of the state of physiological teaching at that time. One finds that Haller's dominating influence was still felt, that Bichat had caused men to modify their conception of tissues, and that the cell theory, though described, was at that date still given cautious commendation".

Some of these notes consist of but a few pages of scrawl and tell little about the lecture contents but, in contrast, those of Lister, taken down in 1849, cover over 400 pages and provide a well-ordered account of what was taught. The basic work of Lavoisier was described and also that of Dutrochet on osmosis and the movement of water. Beaumont's experiments on digestion, using the gastric fistula of Alexis St. Martin, were included complete with tables of digestibility taken from the original, German, edition of Muller's *Handbuch der Physiology* (which did not appear in Baly's English translation). Comparative physiology entered the course with reference to the composition of the blood of lobsters and the embryology of the chick. Although the physico-chemical basis of physiology was becoming to be understood, there was still support for the existence of a special category of animal functions labelled 'vital'; Sharpey was of the opposite, mechanistic school and he used the term 'vital' as meaning only that the actions could not be explained by <u>known</u> physical or chemical laws. As Taylor remarks "A spirit of exactness is now to be seen to pervade physiology". By the standards of the day, Sharpey can be considered advanced; he would have been equally at home with the physiology of a hundred years later.

It was Sharpey's skills as a teacher: his breadth of knowledge, his critical ability, his awareness of experiments, and, above all, his enthusiasm, which made such a profound and lasting impression upon his students.

## PRACTICAL PHYSIOLOGY TEACHING

Although Sharpey had been appointed to teach physiology, this was still mainly by lectures and the only practical work was the microscopic examination of tissues prepared for the students by Sharpey or by his demonstrators. It was a voluntary course, taken by the keenest students. Only one microscope was available and this was passed round the students sitting at a table with a rotating carriage. On rare occasions Sharpey gave a practical demonstration on an animal but there was hardly any student participation. The term 'physiology' meant the examination of the body by instruments or techniques more sophisticated than the scissors or scalpels of the anatomist, foremost among them was the microscope, then looked upon as an ultra-modern instrument not generally accepted by the profession. Thus histology was looked upon as an experimental science, distinct from anatomy, and taught in departments of physiology as an aspect of function, rather than of form. Eventually it was united, some might say re-united, with anatomy in most medical schools.

With the invention of the achromatic lens by Joseph Jackson Lister in 1826, microscopy moved into a new era with greater limits of resolution than ever before. The first systematic account of animal tissues was given by Hodgkin and Lister in 1827; Hodgkin had been in Paris with Sharpey and had supported his application to University College. When in Paris Sharpey may have encountered Alfred Donné who introduced practical histology classes and who wrote a textbook on the subject in 1844. Allen Thomson had also seen the new techniques while studying in Paris and on his return to Edinburgh in 1832 he initiated classes in microscopy for his students, which may have been the first of such classes in Britain. John Hughes Bennett (1812-1875), also a pupil of Donné, when he succeeded Thomson as Professor of the Institutes of Medicine gave a similar course outlined in a book published in 1841. Paget at St. Bartholomews and Bowman at Kings College were both active histologists about this time and Arthur Hill Hassall wrote the first British textbook of human histology in 1846. There was therefore every reason for Sharpey to keep up with modern developments and to see that this aspect of experimental science was taught at University College.

The new attitude to practical physiology was encouraged by the remarks of Augustus de Morgan, Professor of Mathematics at University College, who wrote to Michael Foster in 1853 about the students for the BA degree:

> "I will venture to say that a large majority of those who have passed the examination in physiology know nothing about the interior of the body from their own observation except that blood

follows a cut in the finger...In learning languages words are things; they are the things to be studied, and the student compares the unknown with the known, the strange language with that which he spoken all his life. In the exact sciences the student becomes familiar with the properties of matter, he knows air and water, and his stick is his lever. But in the physiology of the University of London he has only words descriptive of what he neither knows or can know by words alone".

T. H. Huxley expressed the same view when, on a later occasion, he spoke about "the singular unreality of physiological knowledge" which led eventually to an examination requirement by the Royal College of Surgeons and the University as recommended by the General Medical Council in their review of the syllabus in 1866. In January 1855 the Medical Faculty submitted to the Senate proposals for a course on practical physiology made by G. Viner Ellis, (successor to Quain as Professor of Anatomy), Sharpey and William Jenner (Professor of Pathological Anatomy) which had the aim of "supplying to the student of medicine practical instruction in physiology and in the microscopic nature of the textures and fluids of the body" for which a grant of £100 was requested in order to pay for a teacher. In support of their proposal it was pointed out how far University College had fallen behind other medical schools; half of the schools in London gave such courses, so did the schools in Dublin, Scotland and in the English provinces. The *Medical Times and Gazette* published in 1853 a course of ten lectures on histology being given at the Sydenham School of Medicine in Birmingham by a former University College graduate, John Boon Hayes who qualified as MRCS in 1848. He would have been a pupil of Sharpey. and had been a demonstrator with Ellis so he was well known; his name was put forward as someone suitable as a teacher.

Council agreed to the proposal in February 1855 limiting it to two years in the first instance and agreeing to dispense with advertising. Hayes was persuaded to move to London as Lecturer in Practical Physiology and Histology. He gave the first course during that session; according to the University calendar there was an hours class each day and microscopes were provided, an indication that the supply of apparatus had improved since Sharpey's revolving table. However, Hayes resigned in November having obtained an appointment as assistant surgeon to the East India Company; he died in Calcutta from dysentery in July 1856.

The filling of this now vacant post inaugurated the most influential phase of Sharpey's career and a turning point in the development of physiology in England.

The first replacement, in March 1856, was George Harley; he was followed in 1867 by Michael Foster who was succeeded in 1870 by Burdon Sanderson. The post was made permanent and all three were elevated to professorial rank; their careers are examined in detail in the chapter devoted to Sharpey's Men. Under their leadership, practical classes now included chemistry and animal experiments and the position of University College in the physiological world was established.

There is no record of what Harley taught, he was inclined to chemistry rather than to experiments on animals but he carried on with histology courses and, according to the Calendar, he gave a number of physiological demonstrations. In collaboration with his former student George T. Brown MRCVS (later a professor at the Royal Agricultural College, Cirencester) he wrote a manual on *Histological Demonstrations* in 1866. There must have been little or no experimental physiology undertaken by the students since Foster, on his appointment, recalled "What could be done then was very little. I had a small room. I had a few microscopes. But I began to carry out the instruction in a more systematic manner than had ever been done before. For instance, I made the men prepare the tissues for themselves. That was a new thing in histology. And I also made them do for themselves simple experiments on muscles and nerves and other tissues of live animals. That, I may say, was the beginning of the teaching of practical physiology in England".

In Edinburgh, Hughes Bennett had moved on from solely histological teaching to practical experiments on animals for which, as he stated in a letter to the *British Medical Journal* in 1867, he had laboratory space, instruments, and skilled help. His assistant was William Rutherford (1839-1899) who later went to Kings College, London, where he enthusiastically promoted practical physiology for medical students. Bennett made the point that Sharpey attributed the advances in scientific medicine to the new, experimental, outlook fostered by the practical work that was now being done in the schools of physiology.

Under Burdon Sanderson the course at University College expanded still more and by 1873 he was able to publish his *Handbook for the*

*Physiological Laboratory* which described a wide range of procedures. Thus the twenty years following the establishment of the new laboratory saw the growth of physiology as a university discipline and its acceptance as a practical subject in medical education almost as important as anatomy.

Why was it almost twenty years <u>after</u> Sharpey's appointment before the laboratory was founded? In 1836 the expectations for physiology were lower than in 1856; even the use of the microscope was novel. Although Sharpey' held the first chair of physiology in England it was still combined with anatomy and Sharpey was primarily an anatomist all his life. It was not until the 1850's that European physiology burst into flower with the great discoveries of Bernard, du Bois Reymond, Helmholtz and Ludwig.

Sharpey's initiative was a timely response to the changing situation and from University College practical physiology was taken to Cambridge by Foster, to Oxford by Burdon Sanderson and to Edinburgh by Schafer, who reinforced the movement already begun there by Bennett. In a letter to Nature in 1870, the American physiologist H. P. Bowditch wrote "Since in England we have absolutely no physiological laboratory open to students, an account of the best in Germany (Ludwig's) will be interesting. Perhaps some day the University of Oxford will think it desirable to erect such a laboratory to match that recently provided for experimental physics". This brought a response from Burdon Sanderson who was able to point out that such a laboratory had been in existence for many years at University College London and that good work was being done at Edinburgh and elsewhere; the shortage was of physiologists not of laboratories. By the end of the century London was able to challenge the hitherto pre-eminent position of Leipzig.

CHAPTER 5

# *Sharpey's Men*

SHARPEY'S CAREER WAS that of a man who was diligent, painstaking, shrewd, responsible and loyal. He was academically sound but not exceptional and he never became a great scholar, who made major advances in his subject, or a great administrator who founded new organisations or activities. His identification of the perforating fibres of bone was not such as to place him in the first rank of anatomists of the day and he made no physiological discoveries of note. What lifts him above the average, and constitutes the *raison d' etre* of this book, was his ability to impart to some of his pupils and colleagues his enthusiasm and vision, together with his warm, personal friendship, which stimulated them to take up physiology. Their subsequent activities, in teaching and research, brought about a renaissance of physiology in England after a long period of stagnation. They were, in a figurative sense, the real Sharpey's fibres, reaching down into the substance of physiology to make their firm attachments.

These men were talented and would have made their mark in any field but the fact that they chose physiology is deemed to be due, very largely, to Sharpey's inspiration. By its nature evidence for the existence of influence is often sketchy or circumstantial. A person may know in his own mind that he was influenced to move in a particular direction by something said in conversation or by the words or actions of someone personally unknown or even a figure in history. The only evidence of such influence is in the written record of events (nowadays audio-visual records can add to the testimony) and may be expressed in writing only a long time afterwards. Geison (1978) expressed this position very clearly when he stated "there is a need for a more systematic examination of those great teachers who seem to have left behind little in their wake except celebrated students" and later "something could be done to specify the nature and sources of his influence...something more informative than vague tributes of his former students".

I can not claim to have reached these impressive standards but I have examined the writings of or about the men who were pupils or colleagues of Sharpey and who, for all or part of their careers, advanced physiology by their researches or who, in turn, as teachers, inspired a new generation of creative physiologists. Many of his pupils went on to have distinguished careers in other branches of medicine; they may have found him to have been a stimulating teacher but their paths lay elsewhere. It must be remembered that there were very few openings for full-time physiologists; of the nine Sharpey's men considered here only three could be so described. The evidence, then, of Sharpey's influence comes from brief expressions of gratitude in the writings of these physiologists and from similar sentiments by their biographers. None of these disciples wrote autobiographies, often a place where an author can expound upon his early recollections of people and places.

In what follows I have tried to summarise the contributions which each of this select band made to physiology and to give examples of their recognition of Sharpey as a guiding light.

## JAMES BLAKE (1815-1893)

Blake was a student of Sharpey some time before 1841 when he graduated M.B. from University College London. He took up scientific investigation while he was still a student on a subject that arose out of Sharpey's lectures and some of the experiments were done in Sharpey's presence. He used Poiseuille's new mercury manometer to measure the response of blood pressure, and other variables, to the intravenous administration of a number of salts and drugs in dogs and other animals. He was one of the first to investigate the relation between chemical composition and the response of the target organ using the acids and salts of over thirty elements. He had no theoretical basis for his selection of the compounds and no general laws were evolved but his work showed the importance of inorganic salts for mammalian tissues and so anticipated Ringer. This promising work, directly influenced by Sharpey, was not pursued further since Blake emigrated to the United States in 1847 and lived there for most of his life. After a time in practice he became Professor of Obstetrics in San Francisco but he retained an interest in physiology to the extent of publishing four papers in the *Journal of Physiology* between 1884 and 1886 and in 1887 he took the chair at a meeting of the Physiological Society at Kings College, London. He made a lasting contribution to the subject and clearly he can be seen as one of Sharpey's men.

## WILLIAM BALY (1814-1861)

Baly was a student at University College before Sharpey's time, taking the diplomas MRCS and LSA in 1834, but he was connected to Sharpey through the translation of Muller's *Elements of Physiology* into English. In 1836 Baly had taken his MD degree at Berlin and he was fluent in German. He wrote to his father about making the translation and was concerned lest someone else was also doing it. Quain had offered to help him with the plates and to introduce him to the newly appointed Professor Sharpey "who", Baly wrote, "is a very clever man, well acquainted with German literature and might be of great assistance to me". This proved to be correct. As well as giving him valuable advice, Sharpey introduced him to Robert Willis by then the Librarian at the Royal College of Surgeons and author of several books. On another occasion Baly records "I dined with Dr. Allen Thomson of Edinburgh at Dr.Sharpey's". In addition to making these introductions Sharpey also contributed a detailed account of the histology of the placenta based to a large extent upon his own work. The new text was published in 1837-1838 and in it Baly acknowledges the help he had received from Sharpey. It was adopted as a text-book by Sharpey in place of Bostock's but was itself replaced in 1848 by Paget and Kirkes, which was much more in the modern style (it was the original of the long-running *Handbook of Physiology* which, under a succession of authors, went to over forty editions).

Baly took to clinical medicine, not physiology, as a career. He held a number of influential posts including Physician to the Queen but, sadly, he died when only 57 as a result of a railway accident at Wimbledon station.

In memory of his short, but brilliant career a Baly medal was founded at the Royal College of Physicians by Frederick Dyster MD, a general practitioner from Tenby who had no obvious connection with Baly other than qualifying in Berlin (at a later date). In 1866 he offered the sum of £400 to the College to provide a gold medal worth £20 to be awarded in alternate years by the President and Council to the person deemed to have distinguished himself in the science of physiology especially in the two years preceding the award; it was not restricted to British subjects. The third award, in 1873, went to Sharpey. The Baly medallists constitute a hall of fame for physiologists and, as would be expected, Sharpey's men Foster, Burdon Sanderson and Schafer are of their number.

## GEORGE HARLEY (1829-1896)

A fellow Scot by birth, at Haddington, East Lothian, and by education, at Edinburgh where he took his MD in 1846, he emulated Sharpey by choosing science rather than medical practice as a career. To this end he spent the greater part of the next ten years studying in Europe returning occasionally to Edinburgh. In Paris he studied the chemistry of body constituents and attended the lectures of Magendie and Claude Bernard; in Germany he learnt pathology from Virchow, gas analysis from Bunsen in Heidelberg and histology from Sharpey's friend Kolliker in Wurzberg. On his return finally to Edinburgh in 1855 he was offered a post with Hughes Bennett teaching physiology but before he could accept he learnt of the post in Practical Physiology at University College London which had become vacant owing to the sudden resignation of Boon Hayes, as mentioned earlier; he was just in time to apply. He had an informal interview with Sharpey who had heard favourable reports of him from Kolliker, saw in him the right qualities for the job, and gave him the best advice about submitting an application in the short time available.

Harley found himself on a short list of two, the other candidate being Augustus Volney Waller, a very strong contender who had published sound work and was already an FRS. However, Waller withdrew because of ill health leaving the way clear for Harley who was duly appointed. He was an active, ambitious man and in 1859 he combined this lectureship with the chair of Medical Jurisprudence and physician at University College Hospital. He also took up private practice in 1860 thus becoming Dr Harley of Harley Street. He continued to teach practical physiology until 1867 when he was succeeded by Michael Foster. He remained all his life an untiring investigator looking into a wide range of subjects and was elected FRS in 1865. Much of his experimental work was carried out in Sharpey's laboratory.

He remained close to Sharpey and there are several charming references to his friend and mentor in the biography of him written by his daughter Mrs Tweedie. For example, at the first dinner party given by him after their marriage in 1861 the guests included Dr. Sharpey "his patron saint so to speak". Later she wrote that "the five people who most fascinated my father were Liebig, Sharpey, George Catlin (an American traveller and writer) Ruskin and George Waterton the naturalist".

She also recalled a charming anecdote which reflects Sharpey's wry sense of humour:

At luncheon one Sunday Sharpey announced that he was going to be married (he was then over seventy and a confirmed old bachelor). Harley was a bit taken aback but expressed the hope that Sharpey had found a nice sensible, suitable wife who would look after him in his old age. "Nonsense!" said Sharpey, "I am going to marry a most beautiful girl of seventeen or eighteen". Harley was incredulous so Sharpey enlightened him by telling him that for over fifty years, on the advice of his prudent mother, he had subscribed to a Scottish Widows Fund. When he reached the age of sixty and was still unmarried he wrote to the company about discontinuing paying the premium. The company refused to change their rules about a life policy both then and ten years later when he wrote again. Feeling that his days were numbered, he declared that he was ready to marry a young women in order that, on his death, she would be able to secure the pension for which he had paid premiums for over half a century. Mrs. Tweedie concluded "Alas for the sake of the unknown charmer, the good man died unmarried".

George Harley expressed his own very high regard for Sharpey in the following passage:

> "his kindness to rising physiologists is gratefully remembered
> in almost every corner of the globe.....for nineteen years he was
> secretary of the Royal Society and for nearly forty Professor of
> Physiology at University College; thus for years he held the
> destiny of many young men in his hand. A word of generous

encouragement from his lips gave strength to the feeble stem, while one of censure would have nipped the opening bud. But it is to his honour said that, though he never held forth his hand to the undeserving, he never withheld it from those who merited recognition. In no case that I ever heard of did his judgement prove at fault. The false, though it glittered ever so brightly, he never failed to detect. The true coin, were its surface ever so dim, he never failed to recognise".

Harley led an interesting and productive life, giving full reign to his enquiring mind. He wrote two important books, on the chemistry of urine and on liver function, and he was an active laboratory worker on topics such as respiration, the adrenal glands, pancreatic juice and the action of drugs. He contributed many papers to scientific journals and to popular magazines but he is hardly remembered now as a physiologist and, strangely, he never joined the Physiological Society. He certainly advanced the subject by experiment and by teaching and there is ample justification to include him in the select company of Sharpey's men.

## JOSEPH LISTER(1827-1912)

Known throughout the world as the discoverer of the use of antiseptics in surgery, Lister was attracted to physiology by the lectures of Sharpey which he attended as a student at University College London. Although his career was in clinical medicine he nevertheless made a number of valuable contributions to physiology as a form of intellectual recreation in the time that could be spared from practice. In his Huxley lecture of 1900, given when he was an elder statesman of medicine, he recalled that "as a student at University College I was greatly attracted by Dr. Sharpey's lectures, which inspired me with a love of physiology that has never left me".

He was born at Upton, Essex, into a devout Quaker family. His father, Joseph Jackson Lister, a wine shipper and wholesaler in London, was very much interested in science, particularly in optics. He made an outstanding contribution to biological science by his invention of the achromatic lens adding considerably to the resolving power of microscopes, of great importance to the development of histology. His election as FRS in 1832 must have had an influence on his family by making them more familiar with the world of science.

Lister entered University College in 1844 as a student in the Faculty of Arts studying classics and mathematics and graduating BA in 1847. He then entered the medical school where he attended Sharpey's course twice and wrote a set of lecture notes which are still in existence in the College archives. He took the MB degree and the FRCS qualification in 1852 and stayed on at University College Hospital as a house surgeon but he devoted his spare time to physiology. He and Sharpey had developed a strong personal friendship and, on the latter's advice, he left London to become an assistant to Syme, Professor of Surgery at Edinburgh and Sharpey's close friend. Syme had acquired a considerable reputation as a surgeon and Edinburgh held a leading position in the world of medicine so this was an exceptional opportunity for the young Lister to advance his career. During his six years in Edinburgh, in time which could be spared from his expanding surgical practice, he indulged his love of physiology by conducting experiments at his home helped by his new wife, Agnes, the eldest daughter of his chief, James Syme.

In 1860 Lister was appointed to the chair of Surgery at Glasgow where his work on antisepsis started. The move was not as smooth as his growing reputation warranted since, as a Regius chair, a recommendation had to be made by the Home Secretary on behalf of the Crown. A number of Glasgow Members of Parliament had attempted to influence the selection in favour of a Scottish candidate already working in Glasgow, (a similar narrow nationalism had been expressed over the appointments of both Sharpey and Syme to London). But Allen Thomson, by then a distinguished Professor of Anatomy at Glasgow, very much favoured Lister as being a surgeon of proven ability with an excellent scientific background. He and Sharpey exchanged several letters about how to secure this appointment which eventually took place despite the pressures on behalf of the local man.

Lister stayed in Glasgow until 1869 when he succeeded his father-in-law Syme as Professor of Surgery at Edinburgh. He had hoped to move back to London in 1866 when his former chief, Erichsen, transferred from the chair of Surgery at University College Hospital to the Chair of Clinical Surgery. Erichsen himself had attended Sharpey's lectures as a student and after qualifying had, for a short time, taught physiology at Westminster Hospital Medical School. He also served as an inspector under the Cruelty to Animals Act of 1876 which regulated experiments with animals but his physiological attainments are too slight for him to be looked upon as one of Sharpey's men. In the event, Lister was to be disappointed since the post went to John Marshall (1818-1891), then resident assistant surgeon, by a margin of only one vote. Sharpey voted against Lister, probably on the grounds that Marshall, also his friend, was the older man who had already served the hospital for 18 years and he had a good scientific reputation. Lister still had the time and the ability to make further appointments and in fact returned to London in 1877 as Professor of Surgery at Kings College from which he retired in 1893.

Lister received due recognition as a scientist by his election to the Royal Society in 1860, at the comparatively young age of 33, being one of 479 applicants that year for only 15 Fellowships. He served as Foreign Secretary and then as President from 1895 to 1900. He was knighted in 1884 and created a peer, as baron Lister of Lyme Regis, in 1897. In 1902 he was appointed one of the first members of the Order of Merit by Edward VII.

Perhaps at Sharpey's suggestion, Lister's first research was on histology. Using the excellent achromatic microscope given to him by his father, he described the structure of the smooth muscle of the iris and in skin thereby adding to the findings of Kolliker who had shown the existence of these fibres in the wall of the arterioles where they play a part in vasomotor control. This latter topic was investigated by Lister after he had established himself in Edinburgh. He examined the circulation in the frog's web and sometimes in the bat's wing, as used by his former teacher Wharton Jones, observing the aggregation of red blood corpuscles and changes in the dimensions of the arterioles for which he devised a quantitative scale, a considerable innovation. He investigated the importance of the nervous system in bringing about constriction or relaxation of the vessels by severing, in sequence, their nervous connections starting at the brain and working down to the completely isolated limb. He remained convinced that all vasomotor activity was nervous in origin and he explained the changes he observed in isolated vessels as being due to the existence of ganglia surviving in the tissue. Despite the earlier work of Claude Bernard, Brown-Sequard and Augustus Waller, he failed to recognise the importance of sympathetic nerves and so his otherwise sound experiments have never been seen as a major contribution to the subject. He could not, of course, have foretold the existence of endocrine secretions which would have explained the existence of vasomotor activity in the absence of any nervous connections. He then turned to a study of pigment cells in the frog's skin and found that the movement of the pigment granules could be affected by nervous and mechanical stimuli causing them to disperse throughout the cytoplasm, when darkening, or to concentrate around the nucleus leaving the cytoplasm clear. This was a valuable contribution to the knowledge of pigmentary effector systems before anything was known about their hormonal control.

He made a study of blood clotting showing that, as Sharpey had predicted, it depended upon the integrity of the blood vessel wall and not upon the release of ammonia or other chemicals. Damage to the walls resulted in clotting but the blood remained liquid in vessels taken from amputated limbs provided they were undamaged.

In 1858 he made some experiments on the vagus and spinal nerves but, for no accountable reason, he failed to demonstrate slowing of the heart or intestines, which had been shown by others. However, the paper is of interest in that in it he used for the first time in physiology, at Sharpey's suggestion, the word <u>inhibitory</u> system of nerves as a translation of the German word <u>Hemmungsnervensystem.</u>

Lister kept in touch with Sharpey during this period of active physiological research; in a number of letters he discussed his findings and the best way of presenting them. He also put forward his view that medical education should be on the basis of continual assessment, to encourage critical judgement, rather than on annual examinations which relied upon rote learning; Sharpey agreed with him.

Although Lister did not make any major contributions to physiology, his work was of a high standard for the times and his rigorous use of the experimental method served him well in his later work on antiseptics. He was elected an Honorary Member of the Physiological Society in 1892.


MICHAEL FOSTER (1836-1907)

After school in Huntingdon, where his father was in medical practice, Foster went to University College School in London in 1849 and from there to University College in 1852, where his father had also been a medical student. He entered the faculty of arts, not medicine, and graduated BA in classics in 1854. He then turned to medicine, winning a gold medal for anatomy and physiology, and graduated MB in1858; followed by MD in 1859. He attended Sharpey's lecture course twice, the first time while still an arts student. Such was his enthusiasm for physiology that he undertook experiments on the origin of the heart beat in the snail's heart and gave a paper on his findings to the British Association in the same year as he took his MD. This was an interest he was to pursue, alone or with his pupils, for much of his career.

After qualifying he spent some time studying medicine in Paris and then went on a voyage to the Red Sea as a naval surgeon in the hope of studying the natural history of the area. In this he was frustrated and so he resigned his commission and joined his father in practice in his native Huntingdon. From 1860 to 1866 he led the life of a country doctor except that his interest in research remained very much alive and he conducted experiments whenever he found time free from his patients. Some of his work was superficial but he did make the important discovery of glycogen in the body of Ascarid round worms of the pig. His broad understanding of biology led him to speculate about the possible role of this source of energy in an animal which had no requirement to maintain its body tem-perature or to undertake any muscular activity. This work was communi-cated to the Royal Society by Huxley who was an influential figure in Foster's life second only to Sharpey.

Foster must have kept in touch with Sharpey at University College since in 1866 he stepped in to give Harley's lectures on physiology while the latter was afflicted with glaucoma and confined to a darkened room. Harley gave up his post as teacher of Practical Physiology and Foster was invited by Sharpey in January 1869 to apply for the vacant post. There were two other applicants, neither of whom had Foster's experience and they never became known as physiologists. Foster wisely sought an outside reference to support his application and to this end he wrote to Allen Thomson. His letter (number 79 in Jacyna), shows his enthusiasm for the subject: "For the last few years I have been busily engaged in practice in the country....but I have never been able to keep myself from physiology... I have deter-mined to apply (for the post at University College) should I succeed my intention is to devote myself almost entirely to physiology". Thomson's reply is not given but it was presumably very favourable and together with Sharpey's glowing testimonial won the day for him. Two years later he was promoted to professor. In the three years he was at University College he made considerable progress in establishing a course in practical physiology which met the new requirements of both the University and the Royal College of Surgeons. Besides histology, he added to the syllabus elements of biochemistry and experiments on animal preparations making it the most advanced course of its kind in England at that time. He lost no opportunity in getting down to his own research and in the next three years he published papers on the microscope, embedding, enzymes and on the myogenic nature of the heart beat. His status was enhanced by giving the Royal Institution lectures in 1869 choosing the subject of involuntary movement, thus reflecting his continued fascination with the origin of the cardiac impulse. His mentor Huxley was then the Fullerian Professor of Physiology to which office Foster succeeded in the following year. Sadly, his wife died that year leaving him with a son and a daughter to care for and he felt obliged to resign from his posts at University College and the Royal Institution and return to medical practice in Huntingdon.

Prior to this, discussions had been taking place at Trinity College, Cambridge, about broadening the scope of its teaching away from the traditional subjects of classics, mathematics and theology. Physics was one growing subject to be considered but the favoured teacher, William Thomson, later lord Kelvin, was unwilling to move from Glasgow. Physiology's claim was promoted by a number of influential people who were anxious to see that preliminary, now termed pre-clinical, medical education would be based upon science rather than upon the empirical 'arts' of medicine. This group took the view that Cambridge would never be able to sustain a clinical school in opposition to near-by London and that the University should build upon its growing strength in the natural sciences to produce a new type of experimentally minded doctor. This view found expression in a *Lancet* editorial in October 1867 which strongly supported the foundation of a pre-clinical school at Cambridge which would "cease to be a mere appendage of the Church but should be so opened up, and the spirit of it so enlarged, that earnest students of Medicine shall receive the same recognition and the same stimulus to work which are now enjoyed by students of the Thirty-nine Articles". Greater numbers of students "should betake themselves to Oxford and Cambridge to catch the spirit of their culture and perpetuate the fame of the profession of Linacre and Caius and of their universities". The situation had improved so far as chemistry, anatomy and physiology were concerned, but there was a need for more teachers and more scholarships in the natural sciences.

One of the leading protagonists in this group was the Professor of Anatomy, George Humphry. His part in the election of Foster is largely circumstantial but there are a number of slight connections between him and Sharpey which justify a brief account of his life up to this time. He could not be regarded as being one of Sharpey's men, as defined here, but he undoubtedly had an influence on the development of physiology at Cambridge and further afield.

George Murray Humphry (1820-1896) was born and educated in East Anglia and studied medicine at St. Bartholomews where one of his teachers was fellow East Anglian James Paget, a physician with an interest in physiology (he and his colleague Kirkes wrote a notable text-book on the subject). In 1840 Humphry obtained a first class and a gold medal in the 1st MB examination in anatomy and physiology at London University. This was the first year that the university awarded degrees in medicine and also the first year with Sharpey as an examiner. Humphry did not continue with the degree course but qualified as MRCS and LSA in 1842-43. He was then appointed surgeon to Addenbrooke's Hospital, Cambridge, the youngest ever at only 22. Early in his career he started giving lectures to medical students and he also took a Cambridge medical degree himself, MB in 1852 and MD in 1859. He wrote a text-book of anatomy and was elected FRS in1859. In 1866, having previously obtained a College office, he was elected Professor of Human Anatomy and he threw his considerable energy into bringing the basic medical subjects into the orbit of the fast developing Natural Sciences Tripos. He himself taught physiology as well as anatomy (as did Sharpey) and he had used experimental methods in his own research. For his classes on histology he used a rotating microscope table similar to the one used by Sharpey in his early days at University College. These may have been in common use but the coincidence with Sharpey, his former examiner, will readily come to mind.

In 1867 he gave the Address in Physiology at the British Association meeting in Nottingham; it would be nice to record that this owed something to his 'sitting at the feet' of Sharpey and winning his gold medal but it is verbose and dealt only with generalities about form and function. His remark that "the microscope has lately been to physiology much what the steam engine has been to manufacture and transit" was appropriate twenty years before but by 1867 experimental physiology had acquired new techniques, particularly in the field of electrical instrumentation.

Nevertheless, with his experience of physiology and his ambitions for Cambridge it would be likely that he would use his influence to obtain the vacant Praelectorship for physiology. He would have known of developments at University College and seen that Foster (another East Anglian) had all the qualities required to promote physiology at Cambridge but it is not known exactly what Humphry did or whether his outlook was in any way influenced by Sharpey.

A further, tenuous, connection with Sharpey is that both men were founder members of the Physiological Society, and were together at Burdon Sanderson's house on the 31st March 1876 for the inaugural meeting. Humphry resigned the following year but he attended a number of meetings as a guest thus maintaining some connection with the Society. He was by then fully committed to anatomy since Foster had taken over physiology and he himself was editor of the *Journal of Anatomy and Physiology* which, with William Turner, he had founded in 1866. Later he was instrumental in founding the Anatomical Society becoming its first President in 1887.

In the present context then, Humphry can be seen as playing a strong supporting role in establishing the independence and prestige of English physiology; he has been referred to by Geison as 'Foster's ambassador' and, according to his obituarist in the *British Medical Journal*, "he did much to bring about the establishment of the Professorship of Physiology to which Dr. Michael Foster was elected"; he was, in spirit, one of Sharpey's men.

Another possible influence in bringing Foster to Cambridge was Huxley. He knew how much Foster had already achieved and was aware that he was available in near-by Huntingdon. As a fervent campaigner for the extension of biological science he would have no doubt been able to make a strong case for his protégé.

Whatever influences were at work, the result was that Foster was appointed Praelector in Physiology in May 1870. The post is similar to that of Reader but it was a college not a university appointment. Nevertheless, students from other colleges were admitted to the new course, with numbers rising each year, so that Foster and physiology made their impact on the university as a whole. On his appointment he brought in Newell Martin as a demonstrator and offered a similar post to Schafer who, however, preferred to stay at University College.

Thus Foster was brought back to physiology to the great and lasting benefit of the subject and of the University of Cambridge. He was due to take up his new appointment in the Michaelmas term and in preparation for this he embarked upon a tour of the leading European physiological laboratories to see for himself the work they were doing in teaching and research. He took with him, in August 1870, his dear friend Sharpey, then aged 68, who had himself made a similar tour between 1827 and 1829 to prepare for his academic career. Because of the difficulties caused by the outbreak of the Franco-Prussian war in July, they were unable to visit French laboratories but they went to Holland, Germany and Switzerland. In 1880 Foster wrote an account of their journey at the request of Allen Thomson who was gathering material for a biography of his old friend Sharpey who had died earlier that year. Thomson died before completing the book but Foster's memoir was found in the archive at Glasgow University and it is given in full in Note 5. Foster wrote delicately and affectionately about his former teacher who obviously took great delight in re-visiting many of the places where he had studied over forty years previously and where, together, they met many of the leading physiologists of Europe. This charming account of their travels displays the warmth of the relationship between the two men who have had an enduring influence upon British physiology. In writing the memoir Foster not only paid tribute to his teacher but he also provided a sketch of European physiology in which he recognised some of the trends which are still evident today. For example, narrow specialisation which confined a scientist to his own particular interest unaware of much that was relevant to him; the lack of appreciation of historical values as shown by the attitude of some of the scientists they visited to old or unusual anatomical specimens; the race to publish in an increasingly competitive environment. It was a voyage of discovery for Foster and of reminiscence for Sharpey.

Foster's career at Cambridge has been described in detail by Geison (1978) and in a number of other works so that a brief summary is all that is needed here in relation to his position as one of Sharpey's men.

It is generally held that Foster was more an organiser than a laboratory worker but he continued for a time to carry out experiments on the snail's heart. In Geison's view, his familiarity with experiments was vital for the evolution of the supremely successful research of his colleagues even if Foster himself "was no Ludwig". It was this familiarity which contributed to, in Geison's words, "the hitherto mysterious process whereby a 'second-rate' man came to found an undeniably great research school …without himself contributing to it in any monumental or permanent fashion", words that could apply almost equally to Sharpey. Under Foster's guidance Langley, Sheridan Lea, Martin and Gaskell made their distinctive contributions and, in turn, attracted to the Department others whose originality and energy made Cambridge one of the leading schools of physiology in the world. Our present knowledge of the structure and functions of the autonomic nervous system is based particularly on the work of his colleagues Langley and Gaskell.

An indication of the influence Foster had on the Cambridge medical school is provided by an editorial in the *British Medical Journal* in 1879. The editor poured scorn on the then struggling Medical Faculty at Oxford, questioning whether such a faculty really existed or was just a myth. He then launched into a eulogy on Cambridge: "Meantime let us briefly mention, by way of a most brilliant and refreshing contrast (to Oxford), that we have lately had the pleasure of seeing at Cambridge …. a class of nearly a hundred studying physiology under Dr. Michael Foster, with such completeness, and with such aids to higher studies, methods of research, and teaching, and such apparatus and apartments, as are not to be seen elsewhere in England"; anatomy and clinical work also came in for praise. It is clear that Foster had achieved a great deal in the comparatively short time since his appointment, and in an academic climate that did not greatly favour the sciences. In fairness, it must be mentioned that Oxford's position in medicine started to recover when Burdon Sanderson, another of Sharpey's men, was appointed to the newly established Waynflete Chair of Physiology in 1882.

Fosters career after leaving London can be charted by reference to his appointments and his publications. He was a founder member of the Physiological Society in 1876; according to Schafer he was in fact "the true begetter" of the Society even though the initial meeting was held at Burdon Sanderson's house in London. It is strange that despite his unswerving devotion to the Society he never became an Honorary Member, perhaps because he preferred to consider himself as a working member rather than a dignitary. He founded the Journal of Physiology in 1878 and was its editor until 1894 thereby having an influence well beyond his own university. He was elected a Fellow of the Royal Society in 1872 and, in 1881, succeeding Huxley once again, he became its Secretary, remaining in office for an unprecedented period of twenty two years. In 1883 he was appointed the first Professor of Physiology at Cambridge; Huxley was on the panel of electors but it was unlikely that his influence was now needed. He played a leading role in the founding of the International Congresses of Physiology the first of which was held at Bern in 1889; he was later elected as permanent president of the movement. He became increasingly involved with the administration of science in the wider community and served on many government committees concerned with health and education. His activities widened still further when he was elected Member of Parliament (1900 to 1906) for the University of London. He was knighted in 1899.

His publications, other than his papers to scientific journals, show his wide range of interests. He contributed a lengthy chapter on nerve and muscle to Burdon Sanderson's *Handbook of the Physiological Laboratory* (1873) and, with his teaching experience, *A Course of Elementary Practical Physiology* (1876). Soon after he brought out his famous *Textbook of Physiology* (1877) which set new standards in both content and style and went to six editions in England as well as being translated into several other languages. His appreciation of the cultural background of physiology found expression in a biography of *Claude Bernard* (1899) and in his scholarly *Lectures on the History of Physiology during the 16th, 17th and 18th Centuries* (1901). In any comparable work devoted to the nineteenth century Foster would receive the recognition due to him as one of its great leaders; Sharpey's appointment of him proved a decisive factor in the history of physiology in Great Britain and beyond.

## HENRY NEWELL MARTIN (1848-1896)

Perhaps Martin should not be regarded as being one of Sharpey's men since he made no acknowledgement to Sharpey that I have found but he was one of the first to devote his career to physiology as a result of his experience at University College London with Sharpey and Foster. He was born and educated in Northern Ireland and came to London when he was sixteen to become an apprentice to a general practitioner. He had matriculated at University College and studied medicine there while still working in the practice. He came to the notice of Foster who, impressed by his enthusiasm for physiology, appointed him as a demonstrator for the practical courses in Sharpey's department in the years 1868-1870. Martin graduated BSc in 1870 and was offered a scholarship at Christ's College, Cambridge to continue his scientific studies. Foster, knowing of his own imminent appointment to Cambridge, encouraged Martin to accept the offer which he did and in 1873 he took a first class degree in Natural Sciences. He had already qualified in medicine at London, MB in 1871, and had taken up demonstrating in Foster's new course at Cambridge. Now, with a Fellowship at Christ's, he seemed set for a career under Foster when, in 1876, he was appointed to the Chair of Biology at the newly instituted Johns Hopkins University at Baltimore on the recommendation of Foster and Huxley. Before leaving for the USA he attended the founder members meeting of the Physiological Society and he remained a member for the rest of his life. In Baltimore he commenced a very fruitful period of research his most notable being on the isolated heart-lung preparation, acknowledged by Starling as the starting point for his own intensive development of the technique thirty years later. This was the subject of his Croonian lecture to the Royal Society in 1883 and led to his election as FRS in 1885. He was a devoted teacher and he worked strenuously for the establishment, in 1893, of the medical school at Johns Hopkins in which the new Department of Physiology was led by his former pupil W. H. Howell.

Thus Martin handed on the teaching he had received from Sharpey and Foster, who remarked on one occasion "So if I have done nothing more, at all events I sent Henry Newell Martin to America".

## JOHN MARSHALL (1818-1891)

Although not strictly one of Sharpey's men, as the term has been used here, Marshall must have had an influence upon physiology and its role in medical education which can be attributed in part to Sharpey. According to his obituary, "his acquaintance with Dr Sharpey speedily ripened into a close intimacy and a warm friendship which remained unbroken till his death" and he was a trustee of Sharpey's Will.

Marshall entered University College London in 1839, qualified in 1844, and became a curator of the Anatomy Museum under Quain and Sharpey until his appointment as assistant surgeon at University College Hospital in 1848 promoted later to full surgeon. He was elected FRS in 1857 having published several papers on anatomical subjects in the Philosophical Transactions but he was never an experimental physiologist. However, he was Fullerian Professor of Physiology 1862-1865 and in 1867 he wrote a two volume text book *Outlines of Physiology*. Although it was said to be based upon Sharpey's lectures, it was obviously brought up to date since it included a contemporary reference to Burdon Sanderson's work on the connections between respiration and the circulation and a description of Marey's sphygmograph and a kymograph, then very modern instruments. Claude Bernard's work on sympathetic nerves and Brown-Sequard's discovery of reflex vasoconstriction in response to cooling the hand were other examples of his awareness of recent work. He also mentioned some laboratory work by Sharpey on the work of the left ventricle but he does not record when it was done. He wrote "it has been shown by Dr. Sharpey that defibrinated ox blood injected into the thoracic aorta of a dog passes freely back to the veins of the lower limbs; also that if the aorta be tied in the abdomen below the origin of the arteries of the stomach and intestines, the blood still returns along the inferior vena cava" ie. it must pass through either peripheral or hepatic portal capillaries. The pressure employed in these experiments as measured with a haemodynamometer was six inches of mercury, about the same as the dog's left ventricular pressure. This is further evidence that Sharpey did more than histology in the laboratory.

Marshall was a founder member of the Physiological Society in 1876 and he was elected to Honorary Membership in 1882.

## JOHN BURDON SANDERSON (1828-1905)

Of the triumvirate which made up the best known of Sharpey's men, Burdon Sanderson probably had the least degree of personal relationship with Sharpey. He was an Edinburgh graduate, and therefore not taught by Sharpey, and he commenced his physiological experiments before he joined

Sharpey's department and so was not directly influenced by him. But he was committed to the advance of physiology as a discipline and to University College as an institution where that discipline flourished, subjects very close to Sharpey's heart.

Born in Jesmond, just north of Newcastle-upon-Tyne, he was the son of Richard Burdon, a wealthy lawyer and land owner who, in later life, turned to a somewhat austere non-conformism in his religious beliefs. His mother was the daughter of Sir James Sanderson, a former Lord Mayor of London, who stipulated in his Will that whoever should marry his daughter, and inherit the dowry, was to take the name of Sanderson. Therefore Richard Burdon became Richard Sanderson and his son was christened John Scott Sanderson. The middle name came from his paternal grandmother Jane Scott whose two brothers became distinguished law lords. He was educated at home and at the age of eighteen entered Edinburgh University as a medical student where his teachers included Allen Thomson and John Hughes Bennett, both of whom were keen advocates of physiology as the scientific basis of medicine.

He qualified MD in 1851 with a gold medal for his thesis on red blood corpuscles. At about the same time he assumed the name of Burdon, by which he is universally known, but it was not strictly an additional surname to be joined with a hyphen and it was not used in this way by Lady Sanderson in her memoir of her husband.

Like so many medical graduates of the time, he went to Paris to gain further experience and while there he attended the lectures of Claude Bernard but he did not immediately turn towards physiology. On his return to England in 1853 he married and took his first post, as medical registrar to St. Mary's Hospital. Two years later he acquired an additional post as Medical Officer of Health for the Paddington district which he held until 1867. He also held positions at the Royal Brompton Hospital, the Middlesex Hospital Medical School, the short-lived Royal Albert Veterinary School and as a medical inspector to the Privy Council. He must have had a very wide-ranging and busy medical career and he never lost touch with pathology and public health which stemmed from these activities. An example of his experimental approach to practical problems is his part in an investigation of artificial respiration following asphyxia which was instigated by the Royal Medical and Chirurgical Society in 1862. Burdon Sanderson carried out several experiments on dogs and lung volumes were determined in dead bodies using a technique that he devised. It is interesting to note that this subject was studied later by his future colleague and successor Edward Schafer.

Burdon Sanderson became increasingly attracted to the physiological basis of medicine and by 1865 he was conducting, in his own time, experiments on circulation and respiration in man using the sphygmograph. This was more than a dilettante interest since it secured his election to the Royal Society in 1867 and the honour of becoming the Croonian lecturer. He published an account of these studies in a monograph entitled *Handbook of the Sphygmograph* (1865). First described by Marey in 1860 this was one of many recording instruments used by scientifically inclined practitioners. Its clinical value was not very great but for Burdon Sanderson it marked the start of his interest in instrumental techniques in physiology; he was the first in England to use a kymograph, made for him by a metal worker at his private laboratory in Howland Street, near to University College.

Although he was now changing the direction of his work, he had acquired a reputation as an outstanding morbid anatomist and in that role he was asked to carry out the post-mortem examination on the Emperor Napoleon 3rd who died near London on the 9th January 1873.

Possibly through knowing Sharpey at the Royal Society, he was able to carry out some of his expanding experimental work at University College and their association developed into a close friendship. This resulted in his being offered the post of Professor of Practical Physiology when Foster moved to Cambridge in 1870; he immediately resigned his other posts and became a full time physiologist.

Shortly before his appointment he had written to the University of London to draw attention to the possibility that it might loose an endowment which had been made in aid of medical research. By the Will of Thomas Brown of Dublin, who died in 1852, the sum of £22,000 was offered for the establishment of an institution for the investigation and treatment of animal disease. This was to take place within nineteen years of his death otherwise the money would pass to Trinity College, Dublin, for other purposes. That period had nearly expired without any action being taken, partly due to legal reasons since the bequest could not be used for the purchase of land or buildings. In July 1870 Burdon Sanderson wrote to the University with the offer of a sum of £4000 to enable the Brown bequest to be accepted. He was in a position to make such a generous offer since he had a substantial inheritance following the death of his father in 1867. He was looking for a post with laboratory facilities for himself at that time and he expected to be appointed director of the new institute. This offer was not immediately accepted and so, with time running out, he was able to come up with an improved offer of support from John Cunliffe, a city merchant. The University now agreed to proceed and by July 1871 a legal framework was in place with Sharpey as Chairman of the Trustees, a site had been purchased and Burdon Sanderson installed as Professor-Superintendent at a salary of £400 a year. He worked at the Brown Animal Sanatory Institution, as it was named, as well as at University College, until 1876 and he maintained his links with it thereafter becoming chairman from 1901-1903. First-class work in physiology and pathology was carried out at The Brown under a succession of brilliant directors; its significance here lies in its part in the progress of physiology inspired by Sharpey and prosecuted by his protégé Burdon Sanderson.

Within three years of his appointment to University College Burdon Sanderson was able to publish the two volume *Handbook for the Physiological Laboratory* written in conjunction with Foster, Klein and Lauder Brunton. This was a ground-breaking achievement in that it marked the transition from the anatomical physiology of the past to the instrumental physiology of the future; from the study of dead tissues to the study of living animals; from observation to experiment. It was an amazing achievement in so short a time. Its title suggests that there were any number of laboratories ready to make use of such a handbook whereas there was only one in England at that time, at University College. Thus the book anticipated very accurately the future development of physiology. An adverse effect of its publication was that it gave encouragement to the anti-vivisection lobby which denounced the operative procedures which were described. This also had repercussions when the whole subject of experiments on animals came under review by a Government Committee leading to the regulations imposed by the Cruelty to Animals Act of 1876, as told in chapter 3.

In addition to the sections on histology by Klein, muscle and nerve by Foster, and digestion and secretion by Lauder Brunton, the *Handbook* had a section by Burdon Sanderson on blood, circulation, and respiration. This showed his familiarity with the work of the French and German laboratories and his increasing experience with animal procedures and with recording instruments. Among the latter were the induction coil stimulator, myograph, cardiometer and various electro-mechanical devices. A few of these could be obtained in England but many were still to be found only in continental Europe. The *Handbook* can be read today with a sense of familiarity since its methodology and terminology were essentially modern and this must have done much to establish experimental physiology as an independent discipline. Schafer once remarked that English physiology had ceased to exist by 1865 but only a few years later it had started on the way to achieving the dominant position in the world that it subsequently occupied. This was very much what Sharpey had aimed for and the authors expressed their gratitude to him for the encouragement he gave them in a dedication to the *Handbook* which reads:

Dear Dr. Sharpey

To you, who have been for many years the friend of physiologists throughout the world, and who, by your original work, by your teaching, by your generous aid and judicious counsel have been the mainstay of physiology in England, we desire to dedicate this attempt to promote the study of our science.

Accept it as a token of our personal regard, as well as of the high value we set on your life-long labours.

Your devoted Friends,

Michael Foster,

J. Burdon Sanderson,

T. Lauder Brunton,

E. Klein.

On Sharpey's retirement in 1874 Burdon Sanderson was appointed as the first Jodrell Professor of Physiology which combined the two former chairs of Anatomy and Physiology and of Practical Physiology. This was a condition imposed by T. Phillips Jodrell when he made his benefaction of £7000; the aim was to strengthen the post as a full-time one. Schafer was appointed Assistant Professor of Practical Physiology at the same time.

There followed a very active period of research by Burdon Sanderson and his new group of colleagues particularly on the emerging topic of electro-physiology of nerve and muscle. He took up Lippmans capillary electrometer very shortly after its introduction into physiology by Marey in 1876 and used it to demonstrate the action potentials of contracting cardiac and striated muscle and, interestingly, of the leaves of the insectivorous plant Venus's fly trap, brought to his attention by Charles Darwin. His exploration of the time relations of the cardiac cycle led to the demonstration of the human electrocardiogram by A. D. Waller in 1887 and of nerve action potentials by Gotch and Horsley in 1888. His physiological work was recognised by the Royal Society with the award of a second Croonian lecture in 1877 and the Royal Medal in 1883. Amid all this activity Burdon Sanderson found time to gather together his colleagues to found the Physiological Society in May 1876.

In November 1882 Burdon Sanderson resigned from University College to become the first Waynflete Professor of Physiology at Oxford which he held until 1895 when he moved to the Regius Chair of Medicine thus returning once more to the clinical interests of his early days in London. He was created a baronet in 1899 but since he had no children the title became extinct on his death.

Budon Sanderson did not achieve as much for physiology at Oxford as he did at London. He was dogged by a strong anti-vivisection sentiment and by support for the traditional view that medicine at Oxford should be based upon scholarly pursuits rather than upon medical technology which was best left to the provincial schools. He found many obstacles in his path when he wanted to build new laboratories and to modernise the curriculum. He became disheartened and on moving to the Regius Chair of Medicine in 1895 he put more effort into promoting pathology; in his Will he left a bequest for pathology rather than for physiology. Nevertheless, during his time as Waynflete Professor he was able to publish several papers as a physiologist, notably on the on the electrical organs of the skate on which subject he gave a third Croonian lecture to the Royal Society in 1889. His last paper, on the electrical activity of muscle, appeared in the same volume of the *Journal of Physiology* (number18, 1895) as that of Oliver and Schafer on the discovery of the pressor activity of the adrenal gland, a pleasing coincidence which would have appealed to Sharpey.

His contribution to physiology by teaching, administration and research was considerable; he was well and truly one of Sharpey's men.

## SYDNEY RINGER (1835-1910)

His inclusion in the band of Sharpey's men is perhaps less well justified than the others who have been mentioned and he himself made no reference to any influence that Sharpey might have had on him. He was of a later generation, born only the year before Sharpey went to London, but as a medical student he attended Sharpey's lectures on the way to qualifying by diploma in 1859 and by degree in 1860. The rest of his career was spent as a physician at University College Hospital, where Sharpey was a governor, and when the chair of Materia Medica became vacant in 1865 Sharpey wrote to Allen Thomson "We took one of our own people-Sydney Ringer-who being a very able man has done the materia medica work very well, and besides he is one of the most promising clinical men in England and withal a very nice fellow". Like Lister before him, Ringer's keenest interest was reserved for the physiological laboratory to which he diverted any moment that could be spared from his professional duties. While still a student he gave a paper to the Royal Society on the alteration of pitch and sound by conduction through different media, an impressive start to his scientific career, and in 1867 he demonstrated the value of nitroglycerine in the treatment of angina.

Ringer's greatest achievement as a physiologist, which won him worldwide recognition, was his discovery of the importance of mineral salts for maintaining the normal activity of tissues *in vitro*. He used the isolated frog heart for most of his experiments and it was through his acute observation that the need for calcium in isotonic saline solutions was discovered. In an earlier paper he had concluded that only sodium chloride was required but when he repeated the work sometime later he found that the heart soon stopped beating. He found that, by mistake, the earlier saline had been made up with tap water instead of distilled water, as was usual. Analysis of the former showed that it contained traces of calcium salts, derived from its chalky origin, and on adding calcium to distilled water the heart continued to beat normally for long periods; in effect Ringer was correcting a mistake when he made his discovery. Later, potassium was also found to be important. Henceforth the name Ringer's solution was used as a generic term for a range of physiological perfusing fluids.

This work was carried out in the physiological laboratory at University College London in the 1880's, after the death of Sharpey, when Burdon Sanderson and then Schafer were in charge. Ringer published over forty papers on the action of drugs and salts on living tissues; he joined the Physiological Society in 1884 and was elected FRS in 1895. A lecture at the hospital was established in his name by his daughter. Ringer was a reserved, rather austere man who lived for his work and had little time for social activities so it is, perhaps, not surprising that he left no account of his relationship with Sharpey. There is little doubt that his interest was stimulated by his experiences as an undergraduate and by his devotion he advanced the subject, as Sharpey would have wished.

## GEORGE OLIVER (1841-1915)

As a medical student at University College he was much influenced by Sharpey's lectures which he attended before graduating MB in 1865. He went into general practice in Harrogate but he retained a very strong interest in the application of science to medical problems. It became customary for him, after 1890, to give up his practice in winter and move to London or Devonshire where he could indulge his love of physiological research which Sharpey had engendered. He wrote an account of the spa waters of Harrogate, a book on the value of urine testing in clinical practice and another book on pulse testing; this latter displayed his flare for mechanical invention with his new forms of sphygmomanometer, haemocytometer and haemoglobinometer. What marks him out as a Sharpey man is his discovery, jointly with Schafer, of the pressor activity of the adrenal gland which was demonstrated to the Physiological Society in 1894. According to Sherrington, who was present, "Oliver and Schafer gave the most dramatic demonstration I ever remember. They made extracts of adrenal glands and injected a little into the circulation. The effect was so great I thought that the mercury would overflow the blood pressure tube". This discovery was of fundamental importance to the evolution of our knowledge of endocrinology and of sympathetic transmitters The full paper appeared the following year and by 1897 Abel and Crawford in America had isolated adrenaline as the active substance. Oliver and Schafer also found that the posterior pituitary gland, but not other tissues, contained an active pressor agent.

Oliver joined the Physiological Society in 1894 and gave the Croonian lecture of the Royal College of Physicians in 1896 but, somewhat surprisingly, he was not elected to the Royal Society despite his obvious qualifications for the honour.

In 1904 Oliver donated the sum of £2000 to the Royal College of Physicians to enable it "to found and endow a lectureship or prize in memory of the late William Sharpey MD FRS my intention being to promote physiological research by observation and experiment on man himself and to encourage the application of physiological knowledge however acquired to the prevention and cure of disease and the prolongation of life" (his letter to the College). This was accepted but it was thought fitting to recognise Oliver's own achievements and so it was established as the Oliver-Sharpey Lecture. The first was given by Oliver himself in 1904 on the subject of Lymph. The names of later lecturers reads like an Order of Merit of British physiology but the most relevant in the present context is that of Oliver's colleague Schafer, in 1908, and of Schafer's grandson Peter Sharpey-Schafer in 1961. It is pleasing to know that this tribute to Sharpey lives on.

## EDWARD ALBERT SHARPEY-SCHAFER (1850-1935)

Born in London, the third son of J. W. H. Schafer, a city merchant, born in Hamburg but who emigrated to Britain in his early life and became a naturalised citizen. Schafer entered University College London in 1868 where he was taught by Sharpey, Foster and Burdon Sanderson. He won a gold medal and scholarship for anatomy and physiology and in 1871 he became the first Sharpey Scholar which gave him teaching and research experience. He qualified in medicine with the MRCS diploma, not the MB degree, in 1874, the year of Sharpey's retirement, and he was at once appointed Assistant Professor of Practical Physiology. He was elected to the Royal Society in 1878 when he was only 28 years old. From 1878–1881 he was Fullerian Professor and in 1883 he succeeded Burdon Sanderson as Jodrell Professor when the latter moved to Oxford. He remained at University College until 1899 when he took the Chair of Physiology at Edinburgh from which he retired in 1933. He was knighted in 1913.

When Sharpey was appointed to London in 1836 the *Lancet* wondered who was this unknown Scotsman ("man of the north") who was supplanting well-qualified Englishmen. The "balance" was restored when Schafer succeeded Rutherford at Edinburgh. It is understandable that London, as a capital city, should attract men of talent and ambition but what prompted Schafer to leave it for the north? He had an established chair in an environment in which the sciences flourished and from which Sharpey had been disinclined to leave. The salary was much better at Edinburgh, which might

32. The table at University College London, used by Sharpey to demonstrate slides. An arm mounted in the centre carried a microscope which passed from student to student on the circular track.

33. William Sharpey in middle age. (Courtesy of the Boston Medical Library in the Francis A. Countway Library of Medicine, Boston, USA)

*34. George Harley,*
*lithograph 1873.*
*(Courtesy of the Wellcome Library,*
*London)*

*35. Joseph Lister. (Courtesy of*
*the Wellcome Library, London)*

*36. Michael Foster. (Courtesy of the Wellcome Library, London)*

*37. John Scott Burdon Sanderson. (Courtesy of the College Collection, University College London)*

*38. Edward Albert Schafer, 1876. (Courtesy of the Wellcome Library, London)*

*39. Sydney Ringer. (Courtesy of the College Collection, University College London)*

40. *Portrait of Allen Thomson in later years by N. M. Hanhart.* (Courtesy of the Scottish
National Portrait Gallery, Edinburgh)

*41. George Oliver. (Courtesy of the Wellcome Library, London)*

*42. William Sharpey in later life. (Courtesy of the National Institute of Health, Bethesda, USA)*

*43. Oil painting of William Sharpey by J. P. Knight 1870, formerly at University College. (Courtesy of the Wellcome Library, London)*

*44. William Sharpey after retirement. (Courtesy of the College Collection, University College London)*

45. *Bust of William Sharpey by W. H. Thornycroft 1871, plaster cast at University College Hospital, London.*

have been important, and he might have enjoyed the challenge of creating a new centre of excellence for the subject he loved. Is it possible that his veneration for Sharpey the Scotsman played any part in his decision?

His contribution to physiology was massive. One of his first tasks as a Sharpey Scholar was to be a joint editor of the 8th edition of Quain's *Elements of Anatomy* (1876) along with Sharpey and Allen Thomson. His first research project was on the histology of the wing muscles of insects and on the structure of intestinal villi in mammals. In 1878 he studied the nerve network of the jelly fish (Aurelia) and in recognising the discontinuity of nerve fibres he anticipated to some extent the later work of Ramon y Cajal on the neuron doctrine. In 1885 he published his well-known *Essential of Histology* which he saw through twelve editions up to 1929 and which continued to be widely used by medical students until after the sixteenth edition in 1954. He had previously written a small laboratory guide to his histology course at University College (1877) and his commitment to teaching found expression again in 1912 when he published *Experimental Physiology* a guide to the now more extensive practical course which was taught in his new department at Edinburgh. Previous to that he had published between 1898 and 1900 the authoritative advanced *Textbok of Physiology* in two volumes, each of over a thousand pages, based upon contributions from leading authorities; the chapters on blood, nerve cells and the cerebral cortex were written by Schafer himself. It was a statement of what was known at the time and its detailed coverage and wealth of references makes it still an excellent historical introduction. In research he moved for a while to the entirely new subject of cerebral localisation of function on which he had as his colleague Victor Horsley who went on to make a considerable reputation in this field.

In 1894 Schafer performed the experiments which established his name in the annals of physiology. These were his experiments in collaboration with George Oliver in which they discovered the pressor activity of extracts of the adrenal glands, and also of the posterior pituitary gland, thus laying the foundations for the nervous and hormonal control of many physiological functions. He was responsible for introducing the term endocrine for the secretions of the ductless glands and he made a masterly summary of the new subject in his book *The Endocrine Organs* in 1916.

He made an interesting departure from experimental physiology in 1903 when he introduced a new method of artificial respiration in man. The link here with his former colleague Burdon Sanderson has already been pointed out. Many procedures had been suggested, some heroic, for the resuscitation of the apparently dead following submersion or suffocation but they were not very effective. Schafer showed that manual compression of the thorax of the prone subject at the level of the lower ribs brought about expiration followed by inspiration due to the elastic recoil. No apparatus was required and the procedure could continue for as long as operator could do the work. It was simple but effective and it was adopted by many life- saving organisations throughout the world; it was not replaced by mouth-to-mouth resuscitation until 1959 in England.

He and his colleagues continued to be active research workers throughout his time in Edinburgh and his department achieved a high reputation for the quality of its output. In addition, two other activities should be mentioned. The *Journal of Anatomy and Physiology*, founded in 1866, was passing through a lean period especially in physiology, and closure was possible but it was suggested that it might survive if it remained solely an anatomical journal. There was now enough new material to justify a second physiological journal especially one free from the somewhat despotic rule of Langley in Cambridge, then owner and editor of the *Journal of Physiology*, founded in 1878. As a result the *Quarterly Journal of Experimental Physiology* was launched in 1908 with Schafer serving as editor until his retirement in 1933.

Another labour of love that Schafer undertook was the preparation of a *History of the Physiological Society during its First Fifty Years* published in 1926. This is an exceptionally detailed and valuable description of people and events during a period of unprecedented progress in physiology in Britain, a period which ended the stagnancy of the previous 75 years. It was fitting that this record should be written by the last surviving founder member of the Society of which he was elected an Honorary Member in 1926. He died in 1935, the first among equals of Sharpey's men.

Schafer, an outwardly austere man, was deeply aware of his debt to Sharpey for the vision of an intellectually satisfying life based upon physiology, a vision undimmed over a career of more than sixty years. His elder son John was given the middle name of Sharpey out of respect for his former teacher. John joined the navy during the first world war and was killed in action while laying mines from his Q-boat in 1916. Schafer's second son,

Thomas, might have continued the Sharpey tradition since he became a medical student at Cambridge and then at University College Hospital but he left before qualifying in order to join the army and he was killed at the battle of Loos in 1915. The loss of both of his sons must have been devastating to a man who so treasured family life and as an act of remembrance of them and to perpetuate the name of his dear friend and mentor, who had no family of his own, in 1918 he added the name Sharpey to his own surname. The medical register and other documents were duly amended. In his own words, he did this "to emphasise his indebtedness to one who inspired his early work, a great scientist and a staunch friend".

John's son Edward Peter, although born Schafer later assumed the name of Sharpey-Schafer after his grandfather. He took a medical degree from Cambridge and University College Hospital and became Professor of Medicine at St. Thomas's Hospital where, in addition to building a thriving department, he pursued research on the physiology of the peripheral circulation of man. He published several papers in the *Journal of Physiology* on the cardiovascular responses to alterations in breathing thus forging a link with Burdon Sanderson who had made a similar study in 1865. Following his grandfather, he gave the Oliver-Sharpey lecture at the Royal College of Physicians in 1961. Sadly, this gifted physiologist died in 1963 when only 55 thus bringing to an end the Sharpey connection.

## CONCLUSION

The men who have been the subjects of this chapter all made contributions, small or large, to the science of physiology and to them Sharpey gave encouragement, opportunity, his extensive knowledge and the appreciation of physiology as a source of intellectual satisfaction.

The direct link between Sharpey and his pupils came to an end in 1935 with the death of Schafer, almost a hundred years after his appointment as the first full-time physiologist in England. Over this period, and particularly after 1858, physiology developed from being a secondary branch of anatomy to the position of an equal and independent science with its recognised place in medical teaching. Sharpey fostered this growth by keeping alive the glowing embers of its former fire until, through the efforts of his disciples, they burst once more into the flames which illuminated English physiology.

# Postscript: Sharpey – The Man

THE DEATH OF SHARPEY on the 11th of April 1880 brought a number of obituary notices in medical and scientific journals and, at a later date, tributes were paid to him by many of his colleagues. Besides giving an account of the principal events in his life, the authors of these articles all expressed their respect for his many personal qualities and these expressions allows us a glimpse, nothing more, into his character.

The *British Medical Journal* concluded its review of his career by quoting this discerning passage from *Nature*:

> "The qualities which chiefly distinguished him intellectually were the variety of his knowledge, the accuracy of his memory (which he retained to the last without appreciable impairment), and his sound discrimination in all matters of doubt or controversy. To his friends he was endeared by his habitual consideration for the welfare and interests of others, his unwillingness to think ill of those whose conduct he disapproved, and his transparent truthfulness. When it is remembered how large was the circle of his acquaintance, and the number of those who, during the thirty eight years of his professional life, came under his personal influence, we may well moderate our grief at parting with him by reflecting on the good that must have accrued from the life and labours of one in whom so vigorous an understanding was united with so genial and sympathetic a nature".

The *Lancet* extolled his many virtues as scientist, teacher and man of affairs, and concluded that had he been more ambitious, less unselfish and perhaps more combative he might have held a more conspicuous position,

> "but this his friends cannot but think would have been a loss to them and a very doubtful gain to him. His softness of heart, and his generous nature may even in some few instances have led

him to befriend those who were scarcely worthy of his kindness, but the gentleness of disposition, which some may have thought amounted to a fault, was counted by his friends as one of the great charms of his character. While they enjoyed the delights which intimate discourse gave, of chatting familiarly with one who could converse on the most varied subjects with a fullness of information and critical acumen, brightened with a lively sense of humour, they nevertheless felt that the great worth of the man lay in the fact that in times of doubt and uncertainty they could come to him for the surest council, cheering encouragement and solid practical help – a friend whose grand unselfishness knew no speck of meanness".

In his address to University College at the start of the new session, in October 1880, Burdon Sanderson spoke of the loss that was felt by the death of Sharpey:

"I must not dwell longer on the intellectual endowment of the dear friend whom we have lost. We shall no longer have the opportunity of drawing for information on the exhaustless store of his memory, or of profiting by his sage councils. But to those of us who knew the moral excellence of his character, his inflexible devotion to duty, his perfect truthfulness, his unwillingness to think evil of others, his ready sympathy with every noble and generous sentiment – his bright example, though he is gone, will still speak."

Sharpey's death was reported to the Physiological Society at its meting on the 14th October 1880. In view of his outstanding contribution to physiology and his personal reputation, it was a somewhat austere and formal announcement:

"The Secretaries were directed to communicate the following minute to the relatives of the late Dr.Sharpey: That the members of this Society desire to record their sense of the loss they have sustained by the death of a distinguished Honorary Member, the late Dr. Sharpey, an event which has deprived Physiology of a true and devoted follower, and physiological workers of a wise counsellor and generous friend".

The fullest and most illuminating of his obituaries was that by Allen Thomson for the Royal Society, reproduced in part for the Royal Society of Edinburgh. He chronicled the early days in Scotland, the extensive travels abroad and the academic career thereafter; only at the end does he come near to displaying the affection which he felt for his dearest friend:

> "Of the more private features of Dr. Sharpey's life and character it is difficult for those who have been most intimate with him to express their estimate in sufficiently moderate terms. While he was universally admired for the extent and accuracy of his acquirements and respected for the soundness of his judgement, he was not less esteemed and beloved for the gentleness of his disposition, the kindness of his heart, and the geniality of his nature. His powers of memory, naturally good, were carefully cultivated by the systematic turn of his mind and strengthened by exercise. His friends remember with delight the readiness with which, in the course of conversation, he could call up a desiderated quotation, or supply a fact on some doubtful point in history, philosophy or science, or tell humorously some anecdote which was equally apposite or amusing. He had not a single enemy, and he numbered among his friends all those who ever had the advantage of being in his society".

Schafer, the last and most devoted of Sharpey's pupils, in an appreciation of him in 1905, coined the sobriquet by which he has become known: "he has rightly been termed the father of modern physiology in this country" and writing later, now as Edward Sharpey-Schafer, he paid another warm tribute in his *History of the Physiological Society*:

> "Sharpey's personal qualities were such as to inspire his pupils and friends with esteem and affection. He had a gentle nature, a genial disposition and a sound judgement. He was devoted to the interests of the institutions to which he was connected, especially University College and the Royal Society. He spared no pains to promote the advancement of physiology, and was full of encouragement to young men desirous of engaging in original work in that and cognate subjects. He knew and was known to everybody in the world of medicine and science".

An indication of the respect for him held by an average student, one who never took up physiology or became in any sense one of Sharpey's men, lies in the verses composed by William Cobbin who entered University College as a medical student in 1870 and qualified MRCS in 1877. He joined the P&O steamship line as a medical officer but then enlisted to serve with the army in South Africa and was killed in the war against the Zulus in 1879. The verses were sent to the College library by a friend several years later. Although not attaining any poetic heights, they do give an indication of the qualities of the elderly professor which made a deep impression upon his student:

> And first He doth attract my wayward song,
> To whom all praise and all respect belong.
> See, as he comes immersed in thought, he stoops,
> His steps are slow and his left shoulder droops:
> All careful to return the prompt salutes
> Which custom to his Peers and Him deputes.
> A grand old man, respectful to the least,
> A mighty Genius, Natures Great high Priest-
> Who for long years has scanned her ample page,
> Each year more simple, yet each year more sage.
> Vast is his knowledge, unalloyed his mind,
> How gentle his rebuke, his wrath how kind,
> Behold where in the theatre he stands
> And holds his spectacles with careful hands,
> E'en while he chides, he strives to cast the blame
> On his own shoulders – surely very shame
> Should bid the thoughtless students silent be,
> Nor such a teacher vex as good as he.
>
> How deep his grief, how piteous his despair
> When with a well-chalked coat and mournful air,
> His face the type of scientific woe,
> With dreadful emphasis, and accent slow
> He told us, spite of all that he could say,
> That in one paper of last Saturday
> Some student, awful in profanity,
> Wrote: *red corpuscles all had nuclei*!!
> enough – we'll pass him over, three times three,
> With one cheer more, for such a man as He.

(Reproduced from *University College Hospital Magazine*, volume 16, 1882, by kind permission of the Director of Library Services, University College London).

Despite his immense love of learning, Sharpey was no recluse, unlike his colleagues Grant or Wharton Jones; he was a welcome guest at dinners and other social engagements where his good humour and supply of anecdotes made him a lively companion. Allen Thomson recalls that "He was a frequent guest at those delightful little dinner parties given at the house of his old friend Dr. Neil Arnott, another of Arbroath's eminent sons".

Sir Henry Thompson, the famous urologist and socialite, had been a pupil of Sharpey; he related how so-called facts were held up for examination and, if proved wrong, were altered in a subsequent lecture. He remembered him as "a worthy, good-tempered and learned Scotchman enjoying great popularity and influence among the students" who was often a guest at the dinners given by Sir Henry for the members of his distinguished social circle. On such occasions, as Schafer recalls, "he was a genial companion, he overflowed with anecdote – the life and soul of a small dinner party – but a man of thorough business capacity".

To these tributes by those who knew him one can add a few comments based upon the narrative of his life that has been given.

Although he had an English father, he was, and he thought of himself as, a Scot, with a deep love of his family, his home town and his country. He returned frequently and it was his last resting place. Although one must be aware of the dangers of national stereotyping, it can be said that Sharpey possessed many of the characteristics that are associated with the Scots: reliable, conscientious, cautious in decision making, shrewd in judgement, reserved in personal matters and perhaps not greatly imaginative. He was certainly not the proverbial canny Scot where money was concerned. When he was considering the financial aspects of a move to Edinburgh, he wrote to Allen Thomson "I can not save money". He was very generous towards his nephew William Colvill and he did not live extravagantly or own fine property in London and he left comparatively little in his Will.

By his appointment to University College he joined the 'tartan army' of expatriate Scots who found positions of influence in education, law, politics and business in London without loosing their national identity. They might live-out their lives in the Home Counties but they, their children and grandchildren would continue to celebrate Burns night and St. Andrew's day.

On coming to London he would have come into contact with a very wide and varied circle of people. Outside the world of University College there was the University of London Senate with its bishops, politicians and peers of the realm; similarly at the Royal Society and on the various public bodies on which he served. He was unlikely to be over-awed by such company and one can imagine him agreeing with the Scottish saying 'there's nae thing like meeting the great to gie ye a guid opinion o'yursel'. Although essentially a modest man, he probably felt proud of the number of distinctions he had been awarded (Note 6). The good people of Arbroath would no doubt be pleased to know that Henry Sharpey's lad was now an Hon. Member of the Royal Bavarian Academy of Science with a decorated scroll held, along with many others, in the Signal Tower Museum.

In his published correspondence Sharpey rarely gives any opinions on contemporary events; little is known about his views on art, politics or religion. And yet he lived in stirring times. By the time he went to London railways had started to spread across the country and he no longer needed to spend several days on the sea whenever he returned to Scotland. His contemporary, Charles Wheatstone, at King's College, had sent the first telegraphic messages thus opening up the world of telecommunications.

It would be expected that he would welcome and comment upon the discovery of ether as an anaesthetic but his remarks on the subject to Allen Thomson were somewhat muted; in his cautious Scots way he was not prepared to accept anything new before he had all the facts. In his letter of the 23rd December 1846 he wrote:

> "A queer but yet a promising practice has just been tried here of stupefying patients by the inhalation of vapour of Sulphuric Ether in order to render them insensible during surgical operations – Liston cut off a man's thigh in (University College) hospital on Monday (12th December) while the individual was under the influence of Ether – I did not see the operation but I am assured that the man felt nothing. It is true that the leg was whipped off quick enough – but the patient had got back to his bed before he felt any pain. – I suspect however that inhalation of Ether vapour if long continued or repeated at short intervals may in indiscreet hands do mischief by causing bronchitis – at least I am aware of this happening with more than one student who has tried it as a substitute for laughing gas. The use of ether

for the above purpose is I understand a Yankee invention – the suggestor of it a Mr. Bigelow of whose name you must have heard".

This is, as Jacyna remarks in a footnote, "a singularly offhand account of the first British use of anaesthesia in a surgical operation" and Sharpey makes no observations whatever about the discovery of chloroform by his fellow countryman James Simpson in the following year. This does not mean that he was unaware of the importance of these historic discoveries but he was not given to any exuberant expression in his writing, and probably not in his speech either. He used anaesthetics in his own experiments on animals and he appreciated the great advantages they brought to the teaching of practical physiology, as he stated in his evidence to the Royal Commission on vivisection.

Other great events such as the Crimean War (1854), the Indian Mutiny (1857), the publication of *The Origin of Species* (1859), the American Civil War (1861), and the opening of the Suez canal (1869) elicited no comments in his letters to Thomson. Even the closer issue of the acceptance of Elizabeth Garrett onto the Medical Register in 1866 brought no response. It is likely that he would have approved but he had retired before University College accepted women students and the University opened its degrees to them (1878). Thomson rightly observed that reticence was part of his character:

> "Dr. Sharpey was a decided liberal both in politics and religion;
> but though ready at all times to express his views with frank-
> ness and vigour, his gentleness of disposition and knowledge of
> the world made it impossible for him ever to give offence by
> unsuitably obtruding his opinions".

Despite his success, as a professor and at the Royal Society, Sharpey was not ambitious. Financially, he turned down the higher salary of an Edinburgh appointment and he did not undertake lucrative part-time med-ical practice. Academically he did not strive to become a great writer or an influential researcher. Within the university world, he did not aspire to become a vice-chancellor or the head of an Oxbridge college. As an admin-istrator he served the Royal Society well but he did not take office in any of the influential departments of state. It could be said that in being appointed to University College he achieved his ambition, to learn and to teach in the company of scholars.

His lasting memorial is not his portrait, his sculpture bust or his great library, but the Sharpey Physiological Scholarship which makes an enduring connection between him and his College and his subject, to both of which he devoted his life.

*46. West face of the Sharpey family gravestone set up by Sharpey in 1873.*

# *Notes*

## 1. FAMILY TREES

### *Descendants of Henry Sharpey*

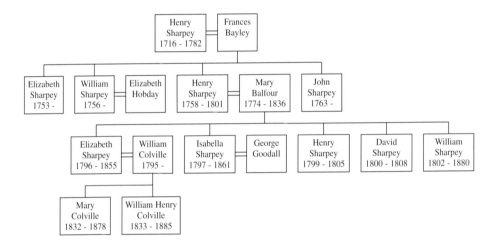

### *Descendants of William Arrott*

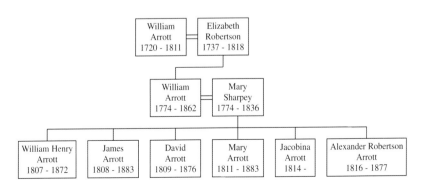

144

## 2. UNIVERSITY COLLEGE COMMITTEES

MEDICAL FACULTY COMMITTEE

*Carswell, Robert* (1793-1857)

Born in Paisley, qualified at Glasgow and Aberdeen, he was a talented artist who studied in Paris before his appointment as professor of Pathological Anatomy in 1828. Such was the quality of his anatomical drawing that he was allowed to return to Paris for three years to complete a series on pathology. His *Illustrations of the Forms of Disease* (1837) is now a classic and a treasure of University College library. Many of his early drawings were made for John Thomson and he was in Paris at the same time as the latter's two sons Allen and William. Carswell retired through ill health in 1840 and went to live in Belgium; he was knighted in1850.

*Davis, David Daniel* (1777- 1841)

Born in Carnarvon, he qualified in Glasgow in 1841. After some time in practice, specialising in midwifery, he was appointed Professor of Obstetrics in 1834. He wrote a number of papers but his chief claim to fame is that he delivered the future Queen Victoria in 1819.

*Grant, Robert Edmond* (1793-1874)

Born in Edinburgh, one of eleven children who all died before him; there were no descendants. He was educated at the High School and at the University from which he qualified as MD in 1814. He studied medicine and natural history in several European cities before returning in 1820 to teach in Edinburgh. It was there that he became interested in marine invertebrates which he found in the tidal waters of the Forth; his studies led to his election as FRSE in 1824. In 1827 he was appointed Professor of Comparative Anatomy and Zoology at the new London University. He was elected FRS in 1836 and was Fullerian Professor at the Royal Institution 1837-1840.He was a conscientious teacher and writer but not an inspiring lecturer; it was said that he never missed giving a lecture in 46 years. His subject was not compulsory for medical students and as a consequence his classes were small. Since the income of a professor depended upon class size, he was often in an impoverished situation. The death of a brother in 1852 brought a legacy which provided for him in some comfort. He could have earned fees from part time practice but he objected to the conditions of the Royal College of Physicians and he was also a martyr to his science; his *magnum opus*, a treatise on comparative anatomy, was never finished. He was one of Sharpey's proposers for election to the Royal Society and they remained on friendly terms throughout their long careers at University College.

*Quain, Richard* (1800-1887)

Born in Ireland and apprenticed to a local practitioner. He came to London as a demonstrator in anatomy, qualified MRCS and was appointed Professor of Anatomy at University College in 1832. For a fuller biography see Chapter 2.

*Thomson, Anthony Todd* (1778-1849)

Born in Edinburgh, he qualified there and moved to London where he was in practice for a time and also took his MRCS. He later added the title of Doctor of Physic from St. Andrews and became FRCP in1842. He was appointed the first Professor of Materia Medica and Therapeutics in 1828 and added to this by becoming Professor of Medical Jurisprudence in 1830. He was an active researcher and wrote many books and articles. His son Alexander was a militant student who played no small part in the campaign which led to the dismissal of Pattison and so to the eventual appointment of Sharpey.

SENATE COMMITTEE

*Carswell, Robert* (as above)

*Elliotson, John* (1791-1868)

Born in London, educated in Edinburgh, where he was President of the Royal Medical Society, and then at Cambridge. He qualified MD in 1821 after study at St. Thomas's and was appointed Professor of the Practice of Medicine in 1831. He was intellectually gifted and very loyal to his hospital but he took up an interest in phrenology and mesmerism and, after some fraudulent seances, he was forced to resign his post in 1838. He had showed his academic qualities with his translation from Latin of Blumenbach's *The Institutions of Physiology* (1815) to which he added copious additions of his own.

*Malden, Henry* (1800-1876)

Chairman. A classicist from Trinity College, Cambridge, he was Professor of Greek from 1831 until he died. For a time he was also joint headmaster of University College School. He was said to be quiet and scholarly, characteristics which he had in common with Sharpey.

*Thomson, Anthony Todd* (as above)

*Turner, Edward* (1798-1887)

Born in Jamaica, he was brought up from an early age in Edinburgh where he qualified as MD in 1818. He was said to be especially interested in anatomy but after study abroad he turned his attention to chemistry and geology. He was appointed the first Professor of Chemistry in 1828 and was elected FRS in 1831. His textbook *Elements of Chemistry* went through eight editions. It was said of him that he was a very careful and painstaking worker rather than an original thinker, qualities he shared with Sharpey.

SENATE

*Booth, James* (1796-1880)

Vice-President of the Senate, he was educated at Cambridge and became a barrister in 1824. He was a council to the Speaker to the House of Commons and later secretary to the Board of Trade.

*Carswell* (as above)

*Cooper, Samuel* (1780-1848)

He qualified MRCS in 1803 after study at St. Bartholomew's Hospital and after practice in the army and in London he was appointed Professor of Surgery to University College in 1834. Elected FRS in 1846. He resigned from his Chair in acrimonious circumstances a few months before his death.

*Davis* (as above)

*Elliotson* (as above)

*Grant* (as above)

*Liston, Robert* (1794-1847)

Born in Scotland, he qualified in Edinburgh in 1816 and spent some time there as surgeon and extra-mural teacher. He was somewhat contentious and was passed over for the chair of Surgery in favour of James Syme. He was appointed to the chair of Clinical Surgery at University College in 1835 and elected FRS in 1841. He was very skilful as an operator and gained lasting fame through his first use of ether as an anaesthetic in England during an amputation on the 21st December 1846.

*Malden* (as above)

*Ritchie, William* (1790-1837)

Born in Scotland, he was originally destined for the Church of Scotland and went so far as to become a licensed preacher with the title of Reverend but he turned to school teaching and was for a time Rector of the Royal Academy at Tain. Having saved enough for the expenses, he moved to Paris to study physics. He returned to London where he lectured and carried out experiments at the Royal Institution. His success there led to his being elected FRS in 1828 and to the Chair of Natural Philosophy at University College in 1832.

*Turner* (as above)

COUNCIL

*Booth, James* (as above)

*Boott, Francis* (no dates found)

MD Edinburgh in 1824 and LRCP in 1826, he settled in practice in Gower Street but also taught at the Webb Street Anatomy School. He was also on the Committee of University College Hospital.

He was born in the USA but brought up in England, and qualified MD Edinburgh in 1824. He lectured in botany at the Webb Street school and became Secretary of the Linean Society.

*Duckworth, Samuel* (no dates found)

*Leader, John Temple* (1810-1903)

Educated at Oxford, he was for a time an MP. A philosophical radical and a London socialite, he left England to live in Cannes and then in Florence as an author and connoisseur.

*Prevost, John Lewis* (no dates found)

*Romilly, John* (1802-1874)
Educated at Cambridge, he entered law at the bar in 1827. He was elected MP and took silk progressing from there to become Solicitor General then Attorney General and finally Master of the Rolls with a peerage.

*Tooke, William* (1777-1863)
A solicitor and a founder member of University College, he was elected FRS in1818 and became MP for Truro.

*Tulk, Charles Augustus* (1786-1858)
Educated at Cambridge, he sat as MP for Sudbury and later for Poole. He devoted himself to good works and intellectual interests.

*Warburton, Henry* (1784-1858)
Educated at Eton and Cambridge, he was one of first members of Council and a philosophical radical in his political views. He was elected FRS and sat as MP for Bridport and then Kendal. He was involved in campaigns for medical reform, the abolition of newspaper duties and for the penny post.

*Waymouth, Henry* (no dates found) Chairman

*Wood, John* (no dates found)
Little is known of him except that he was a civil servant at the Office of Stamps and Taxes.

## 3. ROYAL SOCIETY PROPOSERS

Arnott, Neil (1788-1874) Physician, University of London Senate, author.

Bostock, John (1773-1846) Physician, lecturer, Vice-President Royal Society.

Clift, William (1775-1849) Naturalist, author, Curator of the Hunterian Museum.

Copland, James (1791-1870) Physician, author.

Graham, Thomas (1805-1869) Chemist, Professor at University College.

Grant, Robert Edmond (1793-1874) Anatomist, Professor at University College.

Gray, John Edward (1800-1875) Naturalist, Keeper at British Museum.

Kidd, John (1775-1851) Physician, Regius Professor at Oxford.

Lindley, John (1799-1865) Botanist, Professor at University of London.

Owen, Richard (1804-1892) Anatomist, Keeper at Hunterian and British Museums.

Taylor, John (1779-1863) Mining engineer.

Tooke, William (1777-1863) MP, University of London Senate.

Travers, Benjamin (1783-1858) Surgeon, lecturer, author.

Tweedie, Alexander (1794-1884) Physician, author.

## 4. BIBLIOGRAPHY OF WILLIAM SHARPEY

1.  *De ventriculi carcinomate*, Edinburgh, 1823. Translated as On Cancer of the Stomach, in *Dissertations by Eminent Members of the Royal Medical Society*, Edinburgh, 1892.

2.  *A Probationary Essay on the Pathology and treatment of False Joints*, Edinburgh, 1830.

3.  On a peculiar motion excited in fluids by the surfaces of certain animals, Edinburgh Medical and Surgical Journal, 1830, 34, 113-122.

4.  Remarks on a supposed spontaneous motion of the blood, Edinburgh Journal of Natural and Geographical Science 1831, n. s. II, letter 18 January 1831.

5.  An account of Professor Ehrenberg's more recent researches on the Infusoria, Edinburgh New Philosophical Journal 1833, 15, 287-308.

6.  Observations on the anatomy of the blood vessels of the porpoise, British Association Reports, Edinburgh 1834, 682-683.

7.  Account of the discovery by Purkinje and Valentin of ciliary motion in reptiles and warm blooded animals; with remarks and additional experiments, Edinburgh New Philosophical Journal 1835, 19, 114-129.

8.  Cilia in *Cyclopaedia of Anatomy and Physiology*, edit. R. B. Todd, volume 1, 606-638, London, 1836.

9.  Echinodermata in ibid., vol. 2, 30-46.

10. Anatomy and Physiology - Introductory Lectures, Lancet 1836, 2, 8-88.

11. Anatomy and Physiology – Introductory Lectures, Lancet 1840-1841, 1, 73-77, 142-147, 281-284, 425-428, 489-492.

12. Structure of the decidua and uterine glands. Note by W. Sharpey in *Elements of Physiology*, by J. H. Muller, trans. W. Baly, volume 2, 1574-1582, London, 1842.

13. Cruveilhier, J. *Descriptive Anatomy*, 2 volumes, edit. Alexander Tweedie, revised by W. Sharpey. London, 1841-1842.

14. *Elements of Anatomy*, 5th edition, 1848. Edited by R. Quain and W. Sharpey.

15. *Elements of Anatomy*, 6th edition, 1856. Edited by W. Sharpey and G. V. Ellis.

16. *Elements of Anatomy*, 7th edition, 1864-1867. Edited by W. Sharpey, A. Thomson, and J. Cleland.

17. *Elements of Anatomy*, 8th edition, 1876. Edited by W,. Sharpey, A. Thomson, and E. A. Schafer.

18. The Address in Physiology, British Medical Journal 1862, 2, 162-171.

19. Presidential Address (Biology), British Association, Dundee, 1867, 74-77.

20. Postscript by W. Sharpey in Notes on the Structure of Osseous Tissue, by E. A. Schafer. Quarterly Journal of Microscopical Science, 1878, 18, 132-144.

21. The re-education of the adult brain, Brain, 1879, 2, 1-9.

## 5. FOSTER'S MEMOIR 1880

(Reproduced by permission of The Librarian, Department of Special Collections, Glasgow University Library.)

Sharpey and I left England in August 1870, going first via Harwich to Rotterdam. On our very first day, as throughout the rest of the journey, our interest in the places we visited was largely increased by the extensive and exact knowledge of European history which our friend possessed; this made him in almost every town or city an admirable cicerone.

After a day at La Hague we passed on to Leyden. Here Sharpey was very much interested in Ruysch's preparations in the Anatomical museum. I remember very well his asking to see certain preparations, I forget exactly what, but I think they were of the uterus or foetal membranes; the curator or assistant knew nothing about them; however on searching for them we found them, and Sharpey gave the assistant a lecture on their value and the necessity of taking care of them. Unfortunately all the professors were away. Heynsius's assistant La Place, now professor at Amsterdam, showed us over the place and I remember Sharpey's scorn and his denunciation of the narrowness of modern scientific life when he found that La Place excused himself for taking no interest and knowing little about Teminck's collection, on the grounds of his being a "Physiolog" not an "Ornitholog".

From Leyden we travelled to Amsterdam and after a pleasant day or two there, in which we inspected what was then Kuhne's laboratory, made for Utrecht. Donders was unfortunately away, as also I think Englemann but we saw the laboratory and visited the great church. From Utrecht we went straight to Geissen where we spent a very pleasant day or two with Eckhard, who paid us every attention and in particular showed his erector nerve experiments. I feel uncertain whether Sharpey had been at Geissen before – and yet I fancy he pointed out to me some little gasthaus where he had previously stayed in old times, and which was then one of the principal inns; but he did this in so many places that I have got somewhat confused as to which towns he had not visited.

Leaving Geissen we proceeded via Cassel – where from the rail we had a view of the chateau where Napoleon III was at the time a prisoner* - to Gottingen. Here we met Henle, and he and Sharpey talked over old times and old things with great interest. In particular I remember their laughing much over the impatience and haste of publication of the present young men. Henle told the tale of a zealous inquirer sending a hurried communication to the Centralblat. f. Med. Wis. And then almost immediately telegraphing to withhold it because it was a mistake- but alas too late, the number was printed and distributed.

From Gottingen we passed on via Nordhausen to Leipzig; Sharpey rather wished to go to Berlin but there was an alarm of cholera there and I begged off for I was very much more nervous about such things than he was.

I remember distinctly Sharpey giving me a vivid description of what Leipizg was when he visited it in 1827, and his inability when we arrived there to recognise hardly more than the Pleissenthurm. We spent several very pleasant days at Leipzig being most hospitably entertained by and ciceroned by poor Czermak. We spent much time in Ludwig's laboratory and the Anatomical Museum. Both Ludwig and Ernst von Weber were unfortunately away. Sharpey was particularly anxious to see in the museum there a preparation which Weber used to speak of, (in his lectures I believe) of a brain without olfactory nerves which had been taken from "the chief scavenger of Leipzig"!! The preparation however could not be found – the curator (who was specially studying the nervous system) knew nothing about it, and Sharpey was very disgusted that a so to speak historical specimen had been lost.

From Leipzig we went to Jena, where we had a very pleasant interview with both Haeckel and Gegenbauer, staying at the old inn (Black Bull or Boar) famous for its memories of Luther. After Jena we visited Weimar, where we stayed a day in honour of Goethe and then took train via Eisenach to Frankfurt. Here Sharpey was anxious we should go to the inn he had stayed in former times – The Swanbut it was so altered that he could not recognise it. I remember very well taking a walk with him through the Jews quarter, the only part of the city that seemed to him the same and even that was being largely demolished.

Heidelberg was our next point. Everybody was away but we thoroughly visited the Laboratories and Museums and Sharpey was quite at home except in the new part of the town near the station. One afternoon he insisted on our ascending the Konigstuhl, of the view from which he had a vivid recollection. So we drove to the Molkenkur and walked up from there. The way was long the day was hot, and when we had got three or four parts up Sharpey was visibly distressed. My anxiety about him was not lessened by his replying to my suggestion that we should stop and rest for a while "No it is very important that we should get to the top". However we did get to the top all right, and were soon restored at the "Restauration" with some biscuits and Kirschwasser. The view was disappointing since the day was thick and misty. Sharpey however told me all that I ought to see, and spoke of his previous visits to the top as well as of his rambles in the woods around Heidelberg. We were longer in the ascent than he anticipated and in going down night overtook us and we lost our way in the grounds between the Molken Kur and the schloss. I guided Sharpey in the darkness, through the broken path called I think the Teufelspad – going backwards myself and directing with my hands at each step his foot onto a safe and solid step. At one time I really thought we should spend the night there. As I was looking about alone for a way out I got entangled in the "Maze" belonging to Molken Kur and, though we could hear each others voices, for some time I could not get back to my friend. At last however the music at the schloss, which had been for a long time silent, struck up again; and guided by that we reached he schloss threw ourselves into an Einspanner and were very glad to be safe back at the hotel.

Leaving Heidelberg we journeyed to Stuttgart, picking up Lyon Playfair on the way – Playfair was at dinner very much struck with astonishment, and as I was with envy, at the hearty way in which our friend could drink his Glas Bier.

From Stuttgart we went to Tubingen, which if remember rightly Sharpey had not visited before. We saw Hoppe-Seyler and Luschka, the anatomical museum and Vierordt's laboratory.

Leaving Tubingen we went straight to Schaffhausen. At Tubingen and along the journey Sharpey was for the first time at fault with his German. He could not get along with the patois. Otherwise he talked with great ease with everyone, account being taken of the deafness in one ear and his occasional forgetfulness of a particular word.

After a very pleasant stay for a day or so at Schaffhausen we found ourselves at Zurich – and here as in other places, Sharpey pointed out the old inn – as usual now a broken down old gasthaus – at which he had stayed at the time when it was the principal inn. We had a very pleasant time with Hermann who showed us his laboratory, and at his house we met Fick, on a visit there and several of the other Zurich professors. Our next stay was at Basel, where we had a long chat with His and saw many of his preparations and photographs. Here Sharpey frightened me by an attack of vertigo, similar I fancy to those he had in later years; in this case it was clearly due to Hermann's profuse hospitality coming too soon after a hearty diner at the Hotel du Lac. Leaving Basel we called again at Heidelberg to pick up some things we had ordered there and then went on to Bonn staying at Bingenand going thence down the Rhine. Here we made the acquaintance of Pfluger who showed us his laboratory, Max Schultz was unfortunately away.

Leaving Bonn on Saturday we ran down to Rolandseck and spent the Sunday there. It was a lovely day and I remember still most vividly the enjoyment we had strolling about and chatting with the beautiful Siebengebruge straight before us.

From Rolandseck we turned our faces homewards sleeping a night at Brussels and reaching England via Ostend after an absence of four or five weeks. Sharpey being an excellent sailor and enjoying heartily a little fun at my expense when I found it advisable in the middle of the passage to lie down rather than walk about and talk.

I don't think I ever had a more delightful or a better companion. We never had the slightest difference about anything; even in irritating circumstances as when we lost a train, or missed our luggage there was never a hasty word. In all travelling arrangements after the plan had been discussed between us, he put himself absolutely in my hands, he was as obedient as a child, doing exactly what I had asked so that the slow movements inseparable from his age and confirmities never caused us any trouble. And I need hardly say that all the journey through every step was enlivened and all the tedium driven away by his inexhaustable store of knowledge, and reflections, and stories.

June 22, 1880                                M. Foster

*Note. Foster did not give any exact dates in this memoir but since he left in August, after the declaration of war between France and Prussia on the 15th July, he would have been travelling in Germany when he heard the news of the battle of Sedan (2nd September 1870) when Napoleon III was taken prisoner to the chateau of Wilhelmshoe.


PERSONS MENTIONED

Place names are where the persons were located at the time of Foster's visit; FMRS = Foreign Member of the Royal Society


Czermack, J. N. 1828-1873, Leipzig. Improved the laryngoscope, applied it to phonetics; designed experimental apparatus; sensory physiology.

Donders, F. C. 1818-1829 FMRS, Utrecht. Physiology of the eye, wrote classic account *Anomalies of Accommodation.*

Eckhard, C. 1822-1915, Geissen. Classical study of the erector nerve.

Englemann, T. W. 1843-1909, Utrecht. Established the myogenic nature of cardiac muscle, measured speed of nervous conduction. Hon. Mem. Physiol. Soc.

Fick, A. 1829-1901, Wurzburg. Applied physics to physiology; method for cardiac output (Fick's principle); source of animal heat not only muscle protein.

Gegenbauer, C. 1826-1903 FMRS, Jena. Comparative anatomy and evolutionary morphology.

Haeckel, E. 1834-1919, Jena. Morphology; anatomy of marine invertebrates.

Henle, J. 1809-1885, Gottingen. Histology (loop of Henle), *Handbook of Anatomy*.

Hermann, L. 1838-1914, Zurich. Nitrogen metabolism; wrote a *Textbook of Physiology*; Hon. Mem. Physiol. Soc.

Heynsius, A. 1831-1885, Leyden.

His, W. 1831-1904, Basel. Histology and embryology; constructed the first microtome; (his son discovered the eponymous cardiac fibres).

Hoppe-Seyler, F. 1825-1895, Tubingen. Biochemistry of haemoglobin, application of absorption spectra to blood pigments.

Kuhne, W. 1837-1900 FMRS, Amsterdam. Histology of motor nerve endings; role of visual purple in retina, his monograph *On the Photochemistry of the Retina* 1878, translated by Foster. Hon. Mem. Phys. Soc.

La Place, Leyden.

Ludwig, C. 1816-1895 FMRS, Leipzig. Invented the kymograph and mercury blood gas pump; his laboratory drew students from all the world. Hon. Mem. Phys. Soc.

Luschka, H. 1820-1875, Tubingen. Physiology of the Eustachian tube; *Anatomy of the Larynx* 1873.

Playfair, Lyon, 1819-1898 FRS, London. Chemist, statesman of science, and a government adviser.

Pfluger, E. 1829-1910, Leipzig. Heidelberg. Respiratory activity of tissues, devised a blood gas pump, glycogen metabolism. Editor of the eponymous *Archiv fur Physiologie*. Hon. Mem. Phys. Soc.

Ruysch, F. 1638-1731 FRS, Amsterdam. Master of injection techniques. Discovered valves on lymphatics.

Schultz, Max J. S. 1825-1874, Heidelberg. Studied blood platelets, retinal histology.

Temminck, C. J. 1778-1858, Leyden. Author of famous *Manuel d'Ornithologie*.

Vierordt, K. 1818-1884, Tubingen. Devised and applied a new sphygmograph.

Weber, E. H. 1795-1878 FMRS, Leipzig. Physics of the circulation, pulse waves; with his brother Edouard he discovered vagal inhibition of the heart.

### 6. DEGREES and DISTINCTIONS
(* = document at the Signal Tower Museum, Arbroath)

| 1821 | Diploma Royal College of Surgeons of Edinburgh* |
|------|--------------------------------------------------|
| 1821 | M. D. University of Edinburgh* |
| 1828 | Certificate Universitas Litterae Berolinensis* |
| 1830 | Fellow, Royal College of Surgeons of Edinburgh* |
| 1834 | Fellow, Royal Society of Edinburgh* |
| 1837 | Fellow, Royal Medical and Chirurgical Society |
| 1838 | Fellow, Royal Society of London |
| 1840 | Founder Member of the Royal Microscopical Society |
| 1849 | Medical Society of Sweden* |
| 1850 | Academy of Natural Sciences, Philadelphia* |
| 1851 | Societé de Biologie, Paris* |
| 1853 | Academia Medicina Chirurgica Genoa* |
| 1855 | Honorary Member, Royal Medical Society, Edinburgh |
| 1858 | Physicaler Medicine Gesellschaft, Wurzburg* |
| 1859 | Royal Bavarian Academy of Science* |
| 1860 | Academy of Science, Munich* |
| 1860 | Hon. Doctor of Laws, University of Edinburgh |
| 1862 | Trustee, Hunterian Museum |
| 1862 | Hon.Member, The Odontological Society* |
| 1864 | Fellow, University of London Senate |
| 1865 | Imperial Zoological and Botanical Society of Vienna* |
| 1868 | Royal Academy of Science, Gottingen* |
| 1874 | Emeritus Professor, University of London |
| 1876 | Hon. Member, The Physiological Society |

## 7. SHARPEY SCHOLARS – THE FIRST FIFTY YEARS

| | |
|---|---|
| 1872 | E. A,. Sharpey-Schafer FRS, Professor of Physiology, Edinburgh. |
| 1873 | George Aldridge George. |
| 1874 | William Murrell, lectured on physiology at Westminster Hospital, London. |
| 1875 | Patrick Geddes, Professor of Botany, Dundee; Town Planning Consultant. |
| 1878 | William North, lectured on physiology at University College London. |
| 1881 | Francis Gotch FRS, Professor of Physiology, Oxford. |
| 1882 | William D Haliburton FRS, Professor of Physiology, King's College, London. |
| 1889 | Leonard Hill FRS, Professor of Physiology, The London Hospital. |
| 1891 | J. Herbert Parsons FRS, Ophthalmologist, London. |
| 1896 | Benjamin Moore FRS, Professor of Biochemistry, Oxford. |
| 1897 | Swale Vincent, Professor of Physiology, Middlesex Hospital, London. |
| 1901 | W. A. Osborne, Professor of Physiology, Melbourne, Australia. |
| 1904 | Henry H. Dale FRS, Nobel Laureate, Wellcome Trustee. |
| 1905 | J. M. Hamill, Inspector of Foods, Public Health Department, London. |
| 1907 | F. H. Scott, Professor of Physiology, Minnesota, USA. |
| 1908 | G. C. Mathison, lectured on physiology at Melbourne, Australia. |
| 1911 | C. Lovatt Evans FRS, Professor of Physiology, University College London. |
| 1919 | D. T. Harris, Professor of Physiology, London Hospital Medical College. |
| 1922 | N. B. Dreyer, lectured on physiology in Canada. |
| 1923 | A. S. Parkes FRS, Professor of Reproductive Physiology, Cambridge. |
| 1924 | Richard J. Lythgoe, Reader in Physiology, University College London. |

## 8. NOTES TAKEN BY STUDENTS OF SHARPEY

UCL = Manuscripts Room, University College London

WI = Wellcome Library, London

Anonymous, (1837-1838) Epitome of Physiology, Manuscripta Medica 4. WI

Ballard, E. (1840-1841) MS ADD 286. UCL

Clover, J. T. (1845-1846) Western MS. WI

Lister, J. (1849-1850) Manuscripta Medica 80, Western MS. WI

Potter, J. P. (1836-1837) MS ADD 285, 1-2. UCL

Quain, R. (1837-1838) 2 vols. Yale University, USA

Students Notes (1842-1843) MS ADD 78. UCL

Thane, G. (1867-1868) MS ADD 284. UCL

Tupp, A. E. C. (1860) MS ADD 283. UCL

Whishaw, J. C. (1855) Western MS MSL 17. WI

## 9. IMAGES OF SHARPEY

SCULPTURES
1.  By Peter Slater (1809-1860) of Edinburgh. Referred to by Sharpey in letter 14, 1839 (Jacyna); exhibited at the Royal Scottish Academy 1836 no. 337, plaster, Dr William Sharpey FRCS Lecturer in Anatomy. (not found).
2.  By W. H. Thornycroft RA (1850-1925) marble. Exhibited at the Royal Academy 1872 and presented to University College (not found). Plaster copies, full size, at University College and the Forfar Museum.112 small size plaster replicas produced; one at the Department of Physiology, University College. This was Thornycroft's first commission for a portrait bust, see Elfrida Manning, *Marble and Bronze. The art and life of Hamo Thornycroft*, London 1982.

PAINTINGS and LITHOGRAPHS
1.  By Alexander Blaikley (1816-1903) 1838. Pybus Collection, University of Newcastle upon Tyne.
2.  By T. Bridgford (1812-1878) RHA. National Library of Medicine, Bethesda, USA.
3.  By John Prescott Knight (1803-1881) RA, for University College (not found). Photograph at the Wellcome Library, London.

PHOTOGRAPHS
A number of undated photographs of Sharpey have been located at the sites below; the same photograph is sometimes found in two or more sites.
Barrault & Jerrard 1874 vol. 2. *The Medical Profession in all Countries.*
Francis A. Countway Library of Medicine, Boston, Mass. USA.
McBain, J. M. 1897 *Eminent Arbrothians.*
National Library of Medicine, Bethesda, Maryland, USA.
National Portrait Gallery, London.
Royal Society, London.
Royal Society of Medicine, London.
University College, London.
Wellcome Library, London.

# Selected Bibliography

The sources below contain references to people and events which have been mentioned in this book. Biographical information also came from the *British Medical Journal, The Lancet, The Medical Directory, Munks Roll of the Royal College of Physicians, Plarrs Lives of the Royal College of Surgeons, The Dictionary of National Biography* and *Who Was Who*.

Bellot, H. H., *University College London 1826-1926*, London, 1929.

Bracegirdle, P. *The Establishment of Histology in the Curriculum of London Medical Schools 1826-1886*, PhD Thesis, London University 1996.

Desmond, A., *Huxley. The Devil's Disciple*, London, 1994.

French R. D. *Antivivisection and Medical Science in Victorian Society*, Princeton, 1975.

Geison G. L. *Michael Foster and the Cambridge School of Physiology*, Princeton, 1978.

Godlee, Rickman, *Lord Lister*, London, 1917.

Hall, M. B. *All Scientists Now*, Cambridge, 1984.

Harte, N. B. *The University of London 1836-1986*, London, 1986.

Harte, N. B. and North, J. *The World of University College 1828-1978*, London 1978.

Huelin, G., *King's College London 1828-1878*, London, 1878.

Jacyna, L. S., *A Tale of Three Cities*. Medical History Supplement no. 9, 1989.

LeFanu, W. R. *Robert Willis – physician, librarian, medical historian*. Proceedings of the XXIII International Congress of the History of Medicine, p. 1111-1115, 1974.

McBain J. M. *Eminent Arbroathians*, Arbroath, 1897.

Merrington, R., *University College Hospital: A History*, London. 1976.

Newman, C., *The Evolution of Medical Education in the Nineteenth Century*, Oxford, 1957.

O'Connor, W. J. *Founders of British Physiology 1820-1885*, Manchester, 1988.

O'Connor, W. J. *British Physiologists 1885-1914*, Manchester, 1991.

Poynter, F. N. ed., *The Evolution of Medical Education in Britain*, London, 1966.

Rae, I. *Knox the Anatomist*, Edinburgh, 1964.

Rothschuh, *K. E.*, *History of Physiology*, New York, 1973.

Sanderson G. *Sir John Burdon Sanderson, a Memoir*, Oxford, 1911.

Sharpey-Schafer E. A. *History of the Physiological Society 1876-1926*, Cambridge, 1927.

Sharpey-Schafer Papers, Wellcome Library, London.

Taylor D. W. *The Life and Teaching of William Sharpey*, Medical History 15, 126-153; 241-259, 1971.

Tweedie, Mrs. Alec, *George Harley FRS, The Life of a London Physician*, London, 1899.

Weatherall, M. W. *Gentlemen, Scientists and Doctors. Medicine at Cambridge 1800-1940*, Cambridge, 2000.

Wilson, G. M. *The Brown Animal Sanatory Institution*, Journal of Hygiene, Cambridge, 82, 155-176, 337-352, 502-521; 83, 171-197, 1979.

# Index of Personal Names

Abel, J. J., 131
Abercrombie, J., 26
Allen, John, 14
Allison, W. P., 15, 26, 42
Arnott, James, 18, 47
Arnott, James Moncrief, 47
Arnott, Neil, 18, 47, 59, 101, 140
Arrott, Alexander (half-brother of WS), 3, 5
Arrott, David (half-brother of WS), 3, 5
Arrott, Henry (half-brother of WS), 3
Arrott, Jacobina (half-sister of WS), 3, 5, 55
Arrott, James (half-brother of WS), 3, 4, 5, 8
Arrott, Mary (née Balfour, mother of WS), 5, 35, 39
Arrott, Mary (half-sister of WS), 3, 5
Arrott, William (second husband of Mary Sharpey, née Balfour), 3, 5, 85
Arrott, William (half-brother of WS), 5

Babbage, Charles, 58
Bain, Alexander, 101
Balfour, David (father of Mary Sharpey), 2
Balfour, David (brother of Mary Sharpey), 2
Balfour, Elizabeth (mother of Mary Balfour), 2
Balfour, Mary (mother of WS; see also Mary Arrott), 2
Ballard, E., 90
Ballingall, George (Sir), 26
Baly, William, 89, 102, 109, 110
Barclay, John, 6, 7, 12, 13, 26
Barlow, Thomas (Sir), 100
Beck, Thomas Snow, 63-70
Beaumont, W., 102
Bedford, (Duke of), 15
Bell, Charles (Sir), 19, 21, 31
Bell, Thomas, 73
Bennett, John Hughes, 41, 42, 103, 105, 106, 110, 125
Bennett, Richard, 21
Bernard, Claude, ix, 92, 94, 95, 97, 98, 106, 110, 114, 124, 125
Bichat, F. M. X., 102

Bigelow, G.,142
Blake, James, 108, 109
Blane, Gilbert (Sir), 61
Boot, Francis, 30, 37
Booth, James, 29, 30, 37
Bostock, John, 35, 89, 109
Bowditch, H. P., 106
Bowman, William, 98, 103
Brodie, Benjamin, 65, 67
Brookes, Joshua, 8
Brougham, (Lord), 16, 44
Brown, G. T., 105
Brown, Thomas, 126, 127
Brown-Sequard, C. E., 97, 114, 124
Bruce, Robert (King of Scotland), 2
Brunton, Lauder, 127, 128
Bunsen, Robert Wilhelm, 110
Burdon, Richard (see Sanderson, Richard Burdon), 125
Burke, William, 7, 8, 70
Busk, George, 90
Byron, George (Lord), 57

Caius, John, 18
Cajal, Ramon y, 133
Campbell, Thomas, 16
Carnarvon (Lord), 80
Carpenter, W. B., 18, 89, 97
Carswell, Robert (Sir), 14, 24, 27, 29, 30, 36, 37, 38
Catlin, George, 111
Charles II (King of England), 57
Christie, S. H., 69, 73
Christison, Robert (Sir), 35, 93
Clementi, G., 91
Cobbin, William, 138
Cohnheim, J. F., 60
Colvill, Elizabeth (sister of WS), 55
Colvill, Mary (niece of WS), 2, 55
Colvill, William (brother-in-law of WS), 2, 53
Colvill, William Henry (nephew of WS), 2, 3, 55, 56, 140
Cooper, Samuel, 30, 38, 42-48

159

# Index of Subjects